ISCHAEMIA

preconditioning
and adaptation

UCL Molecular Pathology series

Editor: **D.S. Latchman**
Department of Molecular Pathology, University College London, The Windeyer Building, 46 Cleveland Street, London W1P 6DB

From Genetics to Gene Therapy

Autoimmune Diseases: focus on Sjögren's syndrome

Apoptosis and Cell Cycle Control in Cancer

Ischaemia: preconditioning and adaptation

ISCHAEMIA

preconditioning and adaptation

Michael S. Marber
Department of Cardiology, The Rayne Institute, St Thomas' Hospital, London, UK

Derek M. Yellon
The Hatter Institute, Department of Cardiology, University College Hospital, London, UK

BIOS SCIENTIFIC PUBLISHERS

First published 1996

A CIP catalogue record for this book is available from the British Library.

ISBN 1 859961 60 6

BIOS Scientific Publishers Ltd
9 Newtec Place, Magdalen Road, Oxford OX4 1RE, UK
Tel. +44 (0) 1865 726286. Fax +44 (0) 1865 246823
World-Wide Web home page: http://www.Bookshop.co.uk/BIOS/

DISTRIBUTORS

Australia and New Zealand
DA Information Services
648 Whitehorse Road, Mitcham
Victoria 3132

India
Viva Books Private Limited
4325/3 Ansari Road
Daryaganj
New Delhi 110002

Singapore and South East Asia
Toppan Company (S) PTE Ltd
38 Liu Fang Road, Jurong
Singapore 2262

USA and Canada
BIOS Scientific Publishers
PO Box 605, Herndon
VA 20172-0605

Typeset by Chandos Electronic Publishing, Stanton Harcourt, UK.
Printed and bound in Great Britain by Biddles Ltd, Guildford and King's Lynn.

Contents

Contributors

Baxter, G.F. The Hatter Institute for Cardiovascular Studies, Department of Academic and Clinical Cardiology, University College Hospital, Grafton Way, London WC1E 6AU, UK

Cohen, M.V. Department of Physiology and Medicine, Division of Cardiology, University of South Alabama, Mobile, AL 36688, USA

Connaughton, M. Department of Cardiology, Royal Brompton Hospital, Sydney Street, London SW3 6NP, UK

Downey, J.M. Department of Physiology and Medicine, MSB 3024, University of South Alabama, Mobile, AL 36688, USA

Grover, G.J. Department of Cardiovascular Biochemistry, Bristol-Myers Squibb Pharmaceutical Research Institute, PO Box 4000, Princeton, NJ 08543-4000, USA

Hearse, D.J. Cardiovascular Research, The Rayne Institute, St Thomas' Hospital, London SE1 7EH, UK

Ikonomidis, J.S. Departments of Cardiovascular Surgery and Clinical Biochemistry, The Toronto Hospital, 200 Elizabeth Street, Toronto, Ontario M4N 1V7, Canada

Jennings, R.B. Department of Pathology, Box 3712, Duke University Medical Center, Durham, NC 27710, USA

Kaszala, K. Department of Pharmacology, Albert Szent-Gyorgyi Medical University, Szeged, Hungary

Kloner, R.A. Heart Institute Research Laboratory, Good Samaritan Hospital, 1225 Wilshire Boulevard, Los Angeles, CA 90017, USA

Lawson, C.S. Department of Cardiology, Kent and Sussex Hospital, Mount Ephraim, Tunbridge Wells, Kent TN4 8AT, UK

Li, R.-K. Departments of Cardiovascular Surgery and Clinical Biochemistry, The Toronto Hospital, 200 Elizabeth Street, Toronto, Ontario M4N 1V7, Canada

Marber, M.S. The Department of Cardiology, The United Medical and Dental Schools of Guy's and St Thomas' Hospitals, The Rayne Institute, St Thomas' Hospital, London SE1 7EH, UK

Menasché, P. Department of Cardiovascular Surgery, Hôpital Lariboisière, 2 Rue Ambroise Paré, 75010, Paris, France

Mickle, D.A.G. Departments of Cardiovascular Surgery and Clinical Biochemistry, The Toronto Hospital, 200 Elizabeth Street, Toronto, Ontario M4N 1V7, Canada

Papp, J.G. Department of Pharmacology, Albert Szent-Gyorgyi Medical University, Szeged, Hungary

Parratt, J.R. Department of Physiology and Pharmacology, University of Strathclyde, Glasgow G1 1XW, Scotland, UK

Perrault, L.P. Department of Cardiovascular Surgery, Hôpital Lariboisière, 2 Rue Ambroise Paré, 75010, Paris, France

Rao, V. Departments of Cardiovascular Surgery and Clinical Biochemistry, The Toronto Hospital, 200 Elizabeth Street, Toronto, Ontario M4N 1V7, Canada

Shirai, T. Departments of Cardiovascular Surgery and Clinical Biochemistry, The Toronto Hospital, 200 Elizabeth Street, Toronto, Ontario M4N 1V7, Canada

Tomai, F. Divisione di Cardiochirurgia, Università de Roma Tor Vergata, The European Hospital, Via Portuense 700, 00149 Roma, Italy

Vegh, A. Department of Pharmacology, Albert Szent-Gyorgyi Medical University, Szeged, Hungary

Weisel, R.D. Departments of Cardiovascular Surgery and Clinical Biochemistry, The Toronto Hospital, 200 Elizabeth Street, Toronto, Ontario M4N 1V7, Canada

Yellon, D.M. The Hatter Institute for Cardiovascular Studies, Department of Academic and Clinical Cardiology, University College Hospital, Grafton Way, London WC1E 6AU, UK

Abbreviations

ΣAd	adenine nucleotide pool
ΣNS	total nucleosides and bases
ADO	adenosine
ADP	adenosine diphosphate
AMI	acute myocardial infarction
AOPCP	α,β-methylene adenosine 5′-phosphate
AP	action potential
aPKC	atypical PKC
APNEA	N^6-[2-(4-aminophenyl)ethyl] adenosine
ATP	adenosine triphosphate
BRA	bradykinin
CABG	coronary artery bypass grafting
cAMP	cyclic adenosine monophosphate
CCPA	2-chloro-*N*-cyclopentyladenosine
CFTR	cystic fibrosis transmembrane conductance regulator
cGMP	cyclic guanosine monophosphate
CK	creatine kinase
CK-MB	MB isozyme of creatine kinase
CP	cardioplegia
CPB	cardiopulmonary bypass
cPKC	conventional PKC
DAG	1,2-diacylglycerol
DOG	1, 2-dioctanoyl-*sn*-glycerol
DPCPX	8-cyclopentyl-1,3-dipropylxanthine
EC_{25}	effective concentration for 25% inhibition
ECF	extracellular fluid
ECG	electrocardiogram
FGF	fibroblast growth factor
GTP	guanosine triphosphate
HEP	high energy phosphate
HSP	heat shock protein
HX	hypoxanthine
INO	inosine
IP_3	inositol-1,4,5-triphosphate
I/R	infarcted myocardium as a proportion of risk zone
K_{ATP}	ATP-sensitive potassium channel
L-NAME	N^6-nitro-L-arginine methyl ester

LAD	left anterior descending
LDH	lactate dehydrogenase
LV	left ventricular
MB	cardiac-specific isoform of CK
ND	nucleotidase
NE	norepinephrine
NO	nitric oxide
nPKC	novel PKC
OAG	1-oleyl-2-acetyl glycerol
PBS	phosphate-buffered saline
PC	preconditioning
PCTA	percutaneous transluminal angioplasty
pHi	intracellular pH
PIA	N^6-1-(phenyl-2*R*-isopropyl)adenosine
PKC	protein kinase C
PTCA	percutaneous transluminal coronary angioplasty
R-PIA	N^6-1-(phenyl-2*R*-isopropyl)adenosine
RACKS	PKC-binding protein
SER	smooth endoplasmic reticulum
SOD	superoxide dismutase
SPT	8-*p*-sulphophenyl theophylline
SR	sarcoplasmic reticulum
ST-segment	portion of ECG between S and T wave
SUR	sulphonylurea receptor
SWOP	second window of protection
VEGF	vascular endothelial growth factor
VF	ventricular fibrillation
VPB	ventricular premature beats
VT	ventricular tachycardia
X	xanthine

Preface

The first observation of ischaemic preconditioning 10 years ago has spawned an intensely active experimental and clinical research field. The level of activity within this field ensures such a rapid pace of progress that most published research does not accurately reflect the latest facts and current controversies.

In December 1995 a multidisciplinary symposium was held at University College London on myocardial adaptation and ischaemic preconditioning. This was held as part of the Annual Molecular Pathology Seminar Series at University College London Medical School.

The purpose of the symposium was to provide delegates with information on the most recent advances by allowing researchers from internationally renowned clinical and academic centres at the 'cutting edge' of preconditioning research to present and discuss their thoughts, hypotheses and latest research results.

The organisers of the meeting (now the editors of this book) felt it was both timely and appropriate to prepare an up-to-date text describing the current controversies and issues surrounding ischaemic preconditioning and other forms of adaptation; participants were therefore asked to contribute to this book. In preparing their manuscripts the authors were encouraged to present a personal view from an original stance. The fruits of their labours constitute the chapters within this book.

In keeping with the original symposium, the structure of this book takes the reader through the basic laboratory studies to the current clinical areas where preconditioning and adaptation may find an application.

We believe that this book gives a unique perspective to this exciting research area and would like to thank the authors for their timely and creative contributions.

Michael S. Marber
Derek M. Yellon

1

Overview of preconditioning against lethal cell injury

Robert B. Jennings
Department of Pathology, Box 3712, Duke University Medical Center, Durham, NC 27710, USA

1. Introduction

The discovery that pretreating myocardium with one or more brief episodes of ischaemia protected it transiently against a sustained episode of ischaemia was a surprise to most experimentalists. Murry *et al.* (1986) discovered this phenomenon in 1986 and named it *preconditioning* or *preconditioning myocardium with ischaemia*.

Over the next few years, the preconditioning effect was confirmed as occurring in the dog heart as well as in a number of other species such as the rabbit (Liu *et al.*, 1991), pig (Schulz *et al.*, 1994), rat (Liu and Downey, 1992), ponies (Williams *et al.*, 1991) and probably man (Deutsch *et al.*, 1990). Moreover, the protection against lethal injury exerted by preconditioning was quite strong. In the canine heart, most myocytes exposed to a sustained episode of severe ischaemia of 40–60 min duration survived, whereas virgin myocytes exposed to the same insult died. The original data showing this effect are reproduced in *Figure 1*.

Ischaemia: preconditioning and adaptation, edited by M.S. Marber and D.M. Yellon

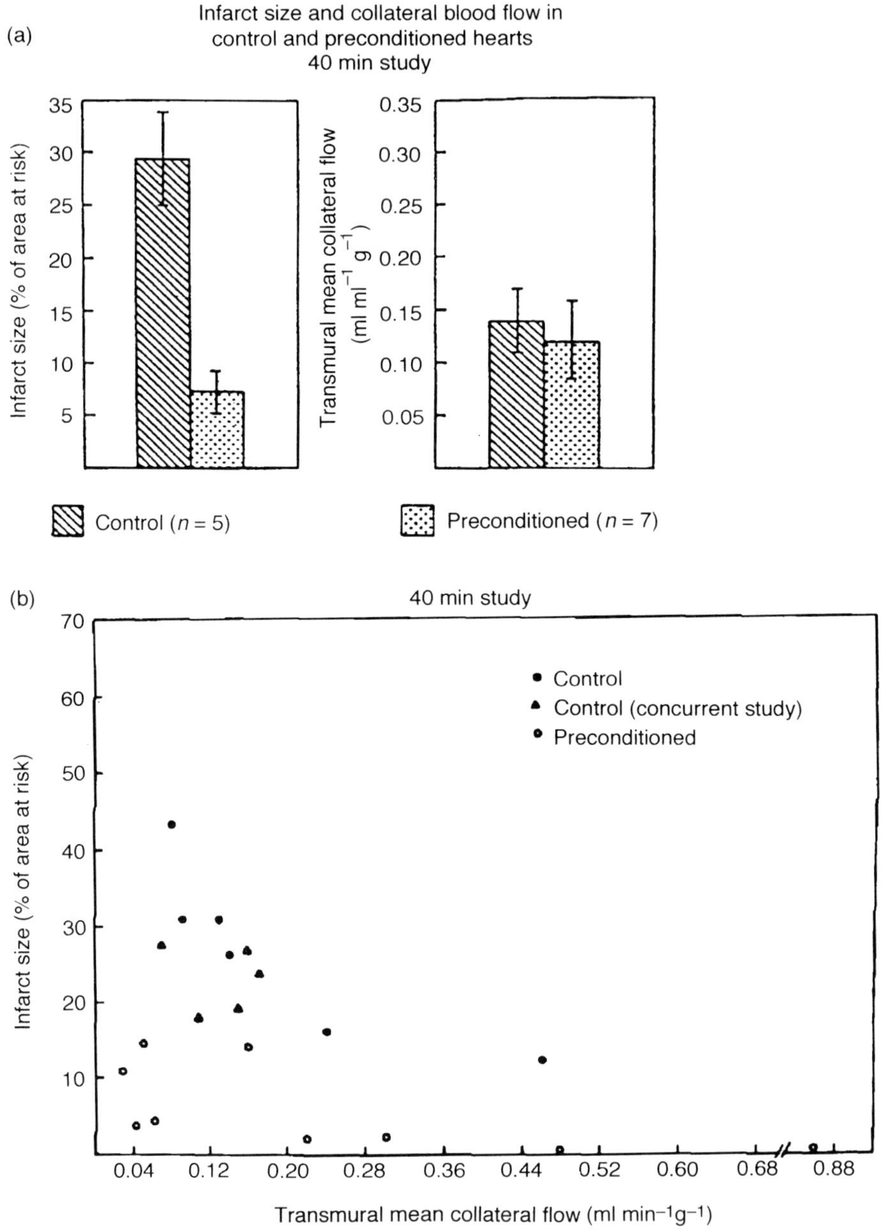

Figure 1. (a) Effect of preconditioning on myocardial infarct size in dogs. Dogs were preconditioned with four 5 min occlusions of the proximal circumflex artery, each separated by 5 min of reperfusion. They were then subjected to a sustained 40-minute ischaemic episode. Control animals received a single 40 min circumflex occlusion. Infarcts were sized histologically after 4 days' reperfusion and were related to the anatomic area at risk and to collateral blood flow (measured with radioactive microspheres midway through the sustained occlusion). In control animals infarct size averaged 29% of the area at risk. In

2. Preconditioning with ischaemia against lethal cell injury.

Over the next 4–5 years, the three major facets of the biology of the preconditioned state with respect to its role in protection against lethal cell injury were established. First, it was shown that the myocardium must be ischaemic for a period of approximately 5 min to precondition (Auchampach and Gross, 1992; Grover *et al.*, 1992). The original preconditioning protocol used four 5-min episodes of ischaemia and reperfusion[a] (Murry *et al.* 1986); however, repeated episodes of ischaemia do not seem to yield significantly greater protection than a single 5 min episode of ischaemia. Five minute episodes of ischaemia also served to precondition the rabbit heart (Liu *et al.*, 1989). Miura *et al.* (1991) have shown that 2 min of ischaemia is adequate to precondition the rabbit heart; however, 5 min of ischaemia is the usual period used by this group. We have found that intervals of ischaemia shorter than 5 min will precondition only an occasional canine heart. Moreover, when the morphine analogue butorphenol (buprenorphine), which has both agonist and antagonist features, is added to the anaesthesia protocol (Schwartz *et al.*, 1995), preconditioning does not occur with a single episode of 5 min of ischaemia.

Less is known about the period of reperfusion required to achieve preconditioning. Most workers have used 5 or 10 min of reperfusion following the episode or episodes of ischaemia. However, it seems likely that a shorter duration would be adequate; our group suspects that the minimum period of reperfusion is that required to restore aerobic

[a] The use of four episodes of ischaemia and reperfusion was derived from a 1985 study on the effect of four 10 min episodes of ischaemia each followed by 20 min of reperfusion on myocardial energy metabolism (Reimer *et al.*, 1986). The results of this study suggested that preconditioning might exist. When we tested the hypothesis, we chose four 5 min episodes of ischaemia and reperfusion in order to shorten the duration of the experiment and because the 10 min ischaemia system was associated with a significant mortality rate from arrhythmias (Murry *et al.*, 1986).

preconditioned animals, infarct size was markedly smaller, averaging 7% of the area at risk. Collateral blood flow to the ischaemic region was not significantly different between groups. (b) Regression of infarct size vs. collateral blood flow. In control animals there was an inverse relation between infarct size and collateral blood flow, that is, low flow was associated with large infarcts and vice versa. In preconditioned animals infarcts were much smaller than controls at any level of collateral blood flow. Reproduced with permission from Murry *et al.* (1986) *Circulation*. © 1986 American Heart Association.

metabolism, and is in keeping with the observations of Alkhulaifi *et al.* (1993).

The second feature of the biology is the fact that the protective effect of ischaemic preconditioning is transient. In the canine heart, it is reduced by about 50% after 120 min of reperfusion (Murry *et al.*, 1991) and is absent after 180 min of reperfusion (Murry *et al.*, 1986). In the rabbit (Miura *et al.*, 1990) and pig heart (Sack *et al.*, 1993) preconditioning is lost after only an hour of reperfusion.

Finally, ischaemic preconditioning is associated with depressed demand for energy during the sustained episode of ischaemia (Murry *et al.*, 1990). This is manifest by slowing in the rate of lactate, H^+ and glycolytic intermediate accumulation and by a slowed rate of adenosine triphosphate (ATP) depletion when compared to virgin myocardium (*Figure* 2). Also, the degradation of the adenine nucleotide pool (ΣAd) and the rate of production of the nucleosides adenosine (ADO) and inosine (INO) is slowed during the first 20 min of sustained ischaemia.

One major feature of the biology of preconditioning is not yet fully established. It is not clear when one can successfully precondition for a second time after preconditioning has disappeared following a prolonged period of reperfusion. Yang *et al.* (1993) found that the rabbit heart could be preconditioned for a second time after preconditioning was lost following 120 min of reperfusion, whereas Schaper's group (Sack *et al.*, 1993) could not reprecondition the pig heart 1 h after the first cycle of preconditioning but were successful after 4 days of reperfusion. Recently, this group (Vogt *et al.*, 1995) has indicated that they can reprecondition a pig heart preconditioned earlier by an episode of ischaemia by infusing ADO into the coronary artery supplying the bed earlier subjected to the preconditioning episode of ischaemia.

Any molecular explanation for the ischaemic preconditioning should be consistent with the foregoing biological features. It should explain: (i) what happens intracellularly during the preconditioning period of ischaemia and reperfusion and how these changes delay the transition to cell death, and (ii) why the protection disappears after 60–120 min of reperfusion, as well as why energy demand is reduced in preconditioned myocardium during the sustained episode of ischaemia.

3. Preconditioning with ischaemia against arrhythmias, stunning and sympathetic denervation

The incidence of arrhythmias is reduced in the isolated perfused rat heart if it is preconditioned with a brief episode of total ischaemia before

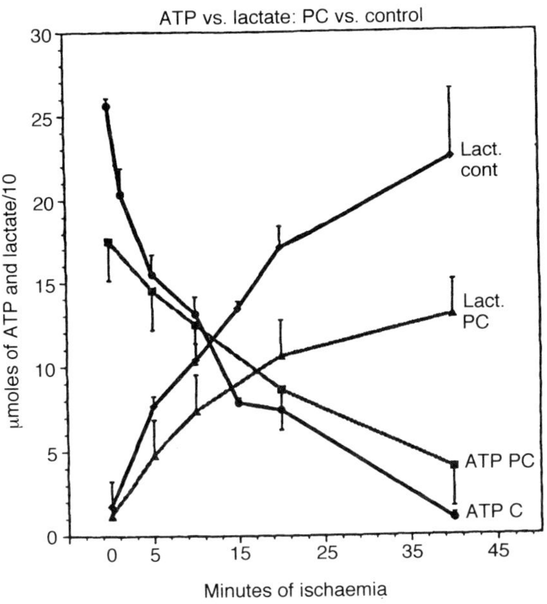

Figure 2. Myocardial ATP and lactate (Lact.) in preconditioned (PC) and virgin heart (C) during 40 minutes of sustained ischaemia induced by proximal occlusion of the circumflex branch of the left coronary artery in dogs anaesthetized with sodium pentobarbital. Each point is the average of three to six hearts with arterial collateral flows of 0.07 ml g^{-1} wet min^{-1} or less and a tissue glycogen greater than 25 μmol glucose equivalents g^{-1} dry. Preconditioning was induced by exposing myocardium of the circumflex bed to four 5 min episodes of ischaemia each followed by reperfusion. The virgin hearts received no pretreatment. All assays were run by enzymatic techniques using neutral perchloric acid extracts prepared from tissue samples that were frozen, lyophilized, and then powdered and assayed. The PC data are from Murry *et al.* (1991). The virgin heart data are from several experiments carried out in our laboratory, using exactly the same procedures and analysts. Some of these data have been reported in another form by Vander Heide *et al.* (1996). The results show accelerated ATP depletion during the first 15 min of ischaemia in the virgin heart vs. a slower steady decline in ATP in the preconditioned tissue. Note that the rate of anaerobic glycolysis as estimated by lactate accumulation, is slower at all intervals in the preconditioned heart. Thus, from the metabolic point of view, PC is characterized by slowed anaerobic glycolysis and slowed depletion of ATP during the sustained episode of ischaemia. Note that tissue ATP declines at essentially similar rates in both preconditioned and virgin heart tissue after 20 min of ischaemia have passed. Reproduced from Jennings and Reimer (1996).

exposing the heart to a sustained episode of ischaemia (Lawson *et al.*, 1993). Vegh *et al.* (1992) have shown a similar protective effect against arrhythmias *in vivo* in the dog heart. It is also clear that the contractile function of the isolated crystalloidal perfused rat heart is improved over that seen in the virgin heart if the heart is preconditioned with a single brief episode of total ischaemia prior to the sustained episode of ischaemia (Cave and Hearse, 1992; Steenbergen *et al.*,1993). The degree of stunning noted in large animal hearts has been reduced in some studies by brief preconditioning episodes of ischaemia (Bunch *et al.*, 1992) but protection against stunning has been generally difficult to demonstrate *in vivo*.

Miyazaki and Zipes (1989) have shown that a brief period of ischaemia has a significant effect on the cardiac sympathetic nervous system. One prolonged episode of reversible ischaemia will effectively denervate the sympathetic nervous system in the ischaemic myocardium. This denervation can be prevented if the heart is exposed to a brief preconditioning episode of ischaemia prior to an episode of sustained ischaemia.

Although unproven, many workers assume that preconditioning with ischaemia induces the same general change in the myocyte that serves to protect it against lethal injury; moreover, it will be impossible to state with certainty that the same underlying mechanism is operative until the molecular mechanism of the preconditioning effect in each model system is established.

4. Delayed preconditioning

Heat stress results in the synthesis of proteins, termed *stress proteins*, that protect cells against ischaemic and other forms of damage. Currie *et al.* (1988) showed that myocardium of animals stressed by heat synthesized increased quantities of these proteins and that such myocardium tolerated ischaemia better than virgin myocardium. These and other studies led to the hypothesis that the protective effect of preconditioning was caused by a genetic response in which stress or related proteins were synthesized during the preconditioning episodes of ischaemia and that these proteins served to protect the cells from ischaemic damage. However, this scenario seemed unlikely because preconditioning developed very quickly, that is, before much gene expression and protein synthesis could occur. For example, one could precondition with 2–5 min of ischaemia and a few minutes of reperfusion, a time period too short to allow significant protein synthesis. Also, Downey's group (Thornton *et al.*, 1990) showed that preconditioning still occurred in hearts in which protein synthesis was inhibited by cycloheximide.

Yellon's group investigated the relationship between stress proteins and protection against lethal injury in detail in the rabbit heart (Yellon and Latchman, 1992; Yellon *et al.*, 1992) and showed clearly that stress proteins were induced when myocardium was preconditioned by ischaemia, so-called *classical preconditioning*, but that these proteins could only be detected many hours later. When present, however, they seemed to be protective, in that the beneficial effect induced by the initial episode of ischaemia, which was lost after an hour, reappeared after prolonged reperfusion presumably because of the presence of these proteins. The new protection was not as strong as that noted during classical preconditioning but clearly was present. They termed it *delayed preconditioning* or the *second window of protection* (SWOP) (Marber *et al.*, 1993). Moreover, the protection persisted for several days rather than the 60–120 min of protection offered by classical preconditioning. Thus, the underlying protective mechanism in SWOP almost certainly is different from that causing classical preconditioning.

5. Preconditioning as an adaptive change

Cells and tissues can adapt to stress by two general mechanisms. The first mechanism is an acute transient form of adaptation brought about by the effect of an agonist released as a consequence of the stress. The agonist binds to a receptor on the target cell and evokes a response. The second is chronic adaptation and involves a genetic response to stress. Here the cell adapts to life in the continuous presence of a stress by net new protein synthesis. Both these general forms of adaptation are found in preconditioning with ischaemia.

5.1. Acute adaptation

β-receptor stimulation of myocytes by norepinephrine (NE) is a classic example of acute adaptation. Within seconds of the release of the NE, it binds to receptors on the myocytes and induces glycogenolysis, increased contractility, and so on. However, the response is short lived; it fades quickly as cyclic AMP, the intracellular enzyme that induces the response, is degraded. A similar response is believed to occur in myocardium secondary to the release of ADO during ischaemia. It binds to A1 receptors on the myocyte and induces the changes of classical preconditioning.

Other potential agonists released during a single episode of ischaemia include bradykinin (BRA) and angiotensin II. These molecules bind to receptors on the myocyte and also mimic preconditioning. Other agents such as endothelin (Hide *et al.*, 1995) and opioids (Chien *et al.*, 1995; Schultz *et al.*, 1995) may be involved. The role of ADO is summarized later in this chapter and indicates that ADO alone can induce an acute adaptive response capable of protecting myocardium transiently from the deleterious effects of ischaemia.

5.2. Chronic adaptation

Cells can adapt to life in the continuous presence of a stress by increased protein synthesis; this form of adaptation is a genetic response to injury that generally persists as long as the stress is present. The best example of chronic adaptation is left-ventricular hypertrophy developing in response to either systemic hypertension or volume overload. In hypertension, for example, the myocytes enlarge, and all components of the myocyte increase in mass, as does the connective tissue and vascular elements of the heart. As a consequence of the myocyte hypertrophy, the left ventricle exhibits increased contractile efficiency. A similar response develops in the liver in mammals exposed to the continuous presence of phenobarbital. Under these circumstances, the smooth endoplasmic reticulum (SER) and the cytochrome P450 enzymes of the SER increase in mass and the organism adapts to the continuous presence of phenobarbital by increasing the capacity of the hepatocytes to degrade phenobarbital.

It seems clear that the stress protein response noted in delayed preconditioning is a similar genetic response. However, it differs from the usual classic chronic adaptive response in that the stress that caused the adaptation is persistent in classic chronic adaptation and disappears in delayed preconditioning. Nevertheless, delayed preconditioning is a response with the same general genetic mechanism as that found in classic chronic adaptation. In all forms of chronic adaptation, the response is considered to be reversible. Eliminate the stress and the tissue tends to return to its original condition.

6. Pharmacologic preconditioning

6.1. Role of adenosine

The penultimate experiment that led to the discovery of preconditioning (Reimer *et al.*, 1986) was a study of the effect of a series of brief episodes of ischaemia and reperfusion on energy metabolism in the canine heart. In this experiment, groups of dog hearts were exposed to one, two or four 10 min episodes of ischaemia each followed by 20 min of reperfusion. The goal of the experiment was the disassociation of ATP depletion from lactate accumulation. The results with the total adenine nucleotide pool (ΣAd), ATP, and nucleosides and bases are shown in *Figure 3*. ATP and ΣAd decreased precipitously during the first 10 min of ischaemia in the virgin heart; the ΣAd pool declined because a large quantity of nucleosides and bases such as ADO, INO, hypoxanthine (HX) and xanthine (X) were formed as a consequence of the fast rate of ~P utilization. Following 20 min of reperfusion, a second episode of ischaemia was associated with a much smaller decrease in ATP and no further loss in ΣAd. Because a significant proportion of the nucleosides and all of the bases are in the extracellular space, at the end of the first 10 min of ischaemia, most of them are lost to the systemic circulation during reperfusion. Since the ΣAd pool essentially was unchanged at the end of the second, third and fourth episodes of ischaemia, the rate of production of nucleosides must have been very small during each of these episodes of ischaemia (*Figure 3*).

The chief source of the nucleosides produced during ischaemia is intracellular adenosine monophosphate (AMP) formed as a consequence of the action of adenylate kinase. AMP is degraded to ADO by intracellular 5′ nucleotidase (Jennings and Steenbergen, 1985). Nucleotides such as AMP cannot exit the living myocyte but nucleosides such as ADO and INO can pass to the extracellular space. Some ADO is degraded intracellularly to INO by ADO deaminase, but clearly much enters the extracellular space where it has been detected in dialysis catheters passing through the ischaemic focus (Van Wylen, 1994). Here ADO is degraded quickly to INO by ADO deaminase in the endothelium of the capillaries. Thus, intracellular adenine nucleotide degradation during the first episode of ischaemia results in the release of much nucleoside to the extracellular fluid in the form of ADO + INO together with their degradation products, the bases HX and X. We found 5 μmol of total nucleoside and bases (ΣNS) at the end of 5 min and 8 μmol at the

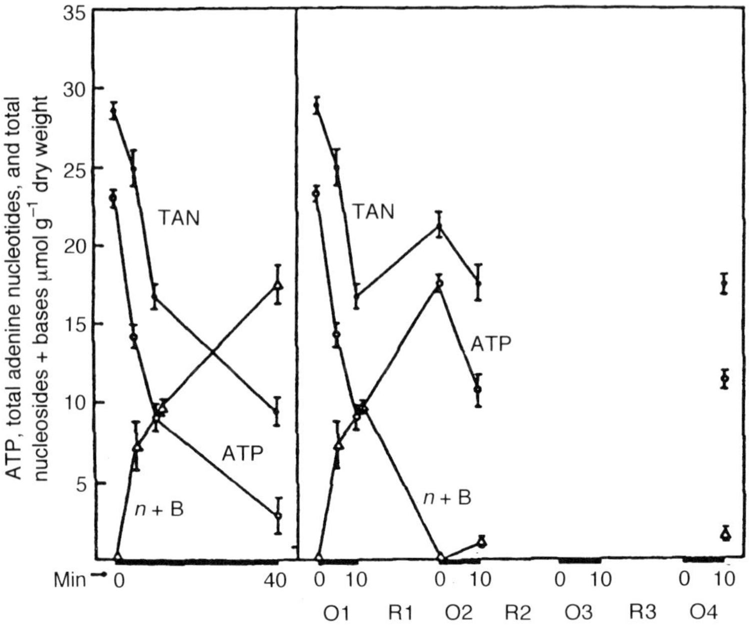

Figure 3. Myocardial adenosine triphosphate (ATP), total adenine nucleotide pool (TAN), and total pool of nucleosides and bases (N + B) during 40 min of continuous ischaemia (left) and during 40 min of cumulative ischaemia with intermittent reperfusion (right). Ten minutes of ischaemia resulted in loss of 60% of the ATP and 40% of the adenine nucleotide pool. Reperfusion for 20 min resulted in only partial repletion of ATP. Nevertheless, no cumulative loss of ATP occurred during three additional 10 min episodes of ischaemia. This is explained by a slower rate of ATP depletion in the later episodes. Note also that production of nucleosides and bases, the catabolites of adenine nucleotides, was substantial in the first ischaemic episode, but was reduced markedly in later episodes. In this experiment, the tissue samples were frozen, dried and sampled without powdering. ATP was measured by enzymatic methods and nucleosides and bases by HPLC techniques. Data presented in *Figure 3* were obtained from frozen dried myocardium that was powdered with a mortar and pestle before sampling. Reproduced from Reimer *et al.* (1986) with permission from the American Physiological Society.

end of 10 min of ischaemia in the virgin heart. In the absence of inhibitors of ADO deaminase or of capillary transport of ADO (dipyridimole), tissue ADO production, as estimated by the serum of ADO + INO + HX + X, never exceeds about 0.8 μmol g^{-1} dry tissue during the second, third or fourth 5 or 10 min episode of ischaemia. Preconditioning with 5 min of ischaemia leads to virtually identical results to those shown in *Figure 3* but the total production of nucleosides is quantitatively smaller.

Some adenine nucleotide can exit from sympathetic neurons during catecholamine release, but this release is believed to be quantitatively small. This nucleotide would be degraded to ADO by the ecto form of 5′ nucleotidase (5′ ND), that is, the 5′ ND of the sarcolemma (Kitakaze *et al.*, 1993).

ADO is a powerful agonist and exerts its effects by binding to A1, A2 and A3 receptors on various myocardial cells including myocytes, vascular smooth muscle cells and sympathetic neurons. The A2 receptor is abundant in blood vessels; when it is stimulated, vasodilatation results. A1 and A3 receptors are present on myocytes; A1 receptors also are present on sympathetic neurons, where stimulation of the receptor inhibits noradrenaline release (Richardt *et al.*, 1989). In the myocytes, stimulation of the A1 receptor results in activation of G_i protein, inhibition of adenyl cyclase and activation of protein kinase C (PKC) (Downey *et al.*, 1994; Ytrehus *et al.*, 1994). Phosphorylation of unknown proteins by the PKC presumably causes the preconditioning effect.

The ADO released during the first episode of ischaemia clearly can bind to A1 receptors and this could be the endogenous mediator that causes preconditioning (Downey *et al.*, 1990; Thornton *et al.*, 1992). Downey's group tested this hypothesis in the rabbit heart using a support rabbit to reduce the systemic effects of ADO on A2 receptors and established unequivocably that one could precondition an otherwise normal heart with a brief infusion of ADO. Furthermore, infusion of N^6-1-(phenyl-2*R*-isopropyl)ADO (R-PIA), an A1 agonist, also successfully preconditioned the heart. Additional support for the role of ADO in preconditioning was provided by the observation that the preconditioning effect could be blocked by 8-*p*-sulphophenyl-theophylline (SPT), an inhibitor of ADO receptor action (Downey *et al.*, 1990). Thus, preconditioning against lethal cell injury was achieved by infusion of a pharmacological agent into virgin myocardium that had never been ischaemic.

It should be noted that pharmacological preconditioning develops in normal myocytes whereas myocytes preconditioned with ischaemia are quite different from control myocytes. They exhibit a variably depressed adenine nucleotide pool, less glycogen and more water (oedema), a CP overshoot (Allison and Holsinger, 1983; Swain *et al.*, 1984), and more P_i and Mg^{2+} (Jennings and Reimer, 1996) than control myocytes. Thus, in pharmacological preconditioning, a change is induced in a normal myocyte with an intact adenine nucleotide pool, that induces the preconditioning effect. To the extent that myocytes preconditioned with ischaemia are damaged as well as being preconditioned, pharmacological preconditioning should delay the transition to cell death to a greater extent than that noted in preconditioning with ischaemia. However, this point has never been specifically addressed experimentally.

These findings with ADO in the rabbit were confirmed in large animal hearts by Grover's group (Grover *et al.*, 1992) who showed that intracoronary R-PIA (an A1 agonist), would successfully precondition the heart against the lethal cell injury. In addition, the beneficial effect of R-PIA and of ischaemic preconditioning both could be blocked by SPT.

The working hypothesis developed from this work is the idea that an endogenous substance such as ADO produced during the first episode of ischaemia stimulates intracellular signalling involving PKC and phosphorylation of proteins by PKC (Louis *et al.*, 1986); it is the phosphorylation or phosphorylations that exert the preconditioning effect. Strong support for this concept was provided by observations in the rabbit heart that staurosporine, chelerythrine and polymyxin B, all relatively non-specific inhibitors of PKC could prevent ADO-mediated and ischaemia-mediated preconditioning (Liu *et al.*, 1994; Ytrehus *et al.*, 1994). However, work in larger animal hearts has been less supportive of this hypothesis (Przyklenk *et al.*, 1995).

Both direct and indirect evidence suggests that sufficient ADO is released during the first episode of ischaemia to precondition the heart. Reinforcement of this hypothesis is provided by the observation that one can speed up the development of the preconditioned state by administration of dipyridamole, an agent that increases the concentration of ADO in the interstitial space (Miura *et al.*, 1992). Unless the receptor has a very high affinity for ADO, ADO is less certain as the agonist during preconditioning induced by repetitive 5 min cycles of ischaemia and reperfusion because much less ADO is available to bind to the A1 receptor during a second 5 min episode of ischaemia. Perhaps the first binding of ADO to the receptor is all that is required to precondition. Alternatively, very little ADO may be required to stimulate the myocyte, or additive effects induced by bradykinin or angiotensin may be involved. The only direct information available is the fact that the quantity of ADO available to stimulate receptors during later episodes of ischaemia clearly is much smaller (*Figure 3*) than that available during the initial episode of ischaemia. The actual events occurring with single and multiple episodes of ischaemia remain unknown.

6.2. Role of other endogenous mediators of preconditioning

Data are available indicating that other endogenous substances are released during ischaemia that could cause or contribute to preconditioning. These include angiotensin II (Liu *et al.*, 1995), endothelin, bradykinin (Martorana *et al.*, 1990; Noda *et al.*, 1993), opioids (Chien *et al.*, 1995; Coker *et al.*, 1981), and so on. In addition, agents such as prostaglandins (Yang *et al.*, 1993) and free radicals (Bolli *et al.*, 1988) are released during reperfusion. These also have been suggested as potential preconditioning agents.

The evidence of the involvement of an endogenous agent in ischaemic preconditioning generally is provided by experiments in which preconditioning with ischaemia is prevented by administration of an inhibitor of the potential agonist such as SPT against ADO (Downey *et al.*, 1990), HOE 140 (Martorana *et al.*, 1990; Noda *et al.*, 1993) against BRA, or enalapril against angiotensin II (Schwarz *et al.*, 1995). Since angiotensin-converting enzyme is a kininase that also degrades BRA, inhibition of this enzyme may increase the concentration of BRA in the area of ischaemia. Also, there is evidence that several of these agents themselves induce pharmacological preconditioning.

7. Role of the K_{ATP} channel

Auchampach and Gross (1992) showed in the canine heart that the ATP-sensitive potassium (K_{ATP}) channel might be involved in preconditioning with ischaemia by demonstrating that one could prevent preconditioning if the hearts were pretreated with the K_{ATP} channel blocker glibenclamide before the preconditioning episode of ischaemia. Their test of glibenclamide was based partly on the observation that A1 receptor activation can open K_{ATP} channels via a G_i protein-mediated process (Kirsch *et al.*, 1990); thus, it seemed likely that the K_{ATP} channels might be involved in preconditioning. In the glibenclamide experiments, dogs were anaesthetized with pentobarbital; the glibenclamide was given intravenously immediately before a 5 min preconditioning episode of ischaemia followed by 10 min of reperfusion. Auchampach and Gross's results were confirmed by Grover *et al.* (Grover *et al.*, 1992) in the dog heart using intracoronary glibenclamide and the same preconditioning protocol. Also, pigs sedated with ketamine and then anaesthetized with thiopental sodium lose preconditioning when given glibenclamide

(Schulz *et al.*, 1994) immediately before the 10 min preconditioning episode and throughout the period of sustained ischaemia. Finally, Grover's group (Grover *et al.*, 1992) showed that pharmacological preconditioning with R-PIA in the dog heart was blocked with glibenclamide. These results taken together indicated probable involvement of the K_{ATP} channel in preconditioning with ischaemia, that ADO might be the mediator of ischaemic preconditioning, and that in both ischaemic and pharmacological preconditioning in large animal hearts, blockade of the K_{ATP} channel with glibenclamide (Grover *et al.*, 1992) essentially eliminated the protective effects of preconditioning against lethal cell injury.

If increased activity of the K_{ATP} channel is involved in preconditioning, then interventions that open this channel should protect myocardium against lethal injury. For example, chromakalin, a K_{ATP} channel opener, will precondition against contractile defects in the isolated perfused rat heart (Grover *et al.*, 1991) and nicorandil, a K_{ATP} channel opener, induces pharmacological preconditioning in the dog heart (Mizumura *et al.*, 1995). These data, together with the glibenclamide data, support the hypothesis that the beneficial effect of preconditioning may be mediated by the K_{ATP} channel.

Direct assessment of K^+ shifts during ischaemia have been made (Hill and Gettes, 1980) using K^+ and H^+ electrodes placed in the centre of an area of *acute* ischaemia. Fleet *et al.* (1985) have studied the effect of episodes of ischaemia and reperfusion on this process and have shown that K^+ in the extracellular fluid (ECF) rises by 2 mmol l^{-1} after only 2 min of ischaemia and that its level is double that found in the plasma after 4 min of ischaemia, at which time it essentially plateaus at 8–9 mmol l^{-1}. Meanwhile, tissue pH has decreased to <6.0. If, however, the heart is preconditioned by 10 min of ischaemia and 50 min of reperfusion and then is exposed to a second 10 min episode of ischaemia, the K^+ of the ECF triples rather than doubles; simultaneously, much less acidosis develops (Fleet *et al.*, 1985). These data are shown in *Figure 4*. Although this phenomenon was described before preconditioning was defined, it clearly shows that more intracellular K^+ exits in preconditioned than in virgin hearts. The decrease in the degree of acidosis developing with preconditioned pig heart (Fleet *et al.*, 1985) is consistent with the low rate of lactate accumulation shown in *Figure 2*.

The hypothetical changes occurring in preconditioning are shown diagrammatically in *Figure 5*. Ischaemia clearly opens K channels, including presumably the K_{ATP} channel, and allows K^+ to accumulate in the ECF. The excess K^+ in the ECF has a cardioplegic effect and

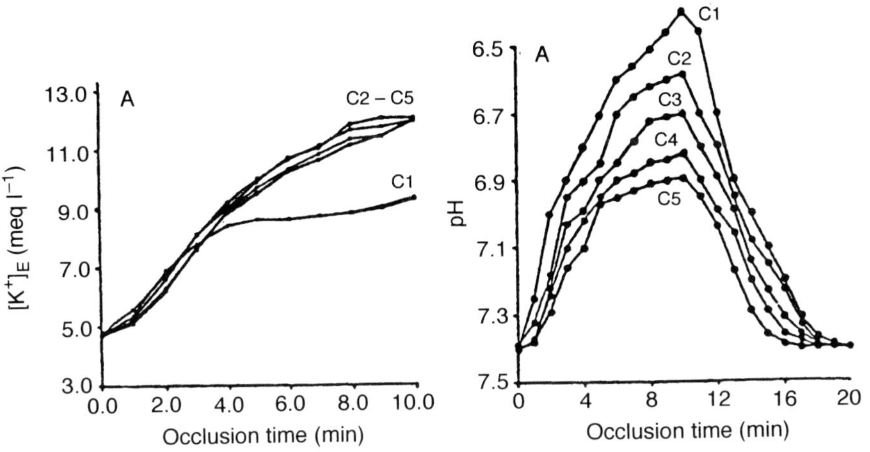

Figure 4. Shifts in K^+ and pH are shown in the extracellular fluid in the inner myocardium in the centre of the zone of ischaemia in acutely ischaemic pig heart. In these panels, C1–C5 are tracings made in the centre of the area of ischaemia during the first to the fifth 10 min episode of ischaemia, each separated by 50 min of reperfusion. See text for description. Reproduced with permission from Fleet *et al.* (1985) *Circulation*. © 1985 American Heart Association.

depolarizes the myocyte. Some of this K^+ exit may be volume regulated (Jennings *et al.*, 1986). Irrespective of this, the increased K^+ release into the ECF acts as a cardioplegic agent in the preconditioned myocardium and thereby reduces energy demand.

Although K_{ATP} channels are abundant in the sarcolemma (Lederer and Nichols, 1995) and contain amino acids that could be phosphorylated readily by PKC, direct demonstration of phosphorylation and the effects of phosphorylation on K_{ATP} channel function have not been described.

It seems likely that the K channel may be the final mediator of the preconditioning effect. The cardioplegic effect of the excess extracellular K^+ in the zone of ischaemia would explain the decreased energy demand. Moreover, the effect would be transient, because the onset of irreversible injury is associated with an even greater increase in extracellular K^+ in both preconditioned and virgin tissue (Hill and Gettes, 1980); at this time, the difference between virgin and preconditioned tissue should and in fact does disappear. In addition if phosphorylation of the channel by activated PKC resulted in persistent opening of the channel, it would help to explain the effects of ADO and PKC inhibitors on preconditioning. Note that deactivation of the channel could follow dephosphorylation.

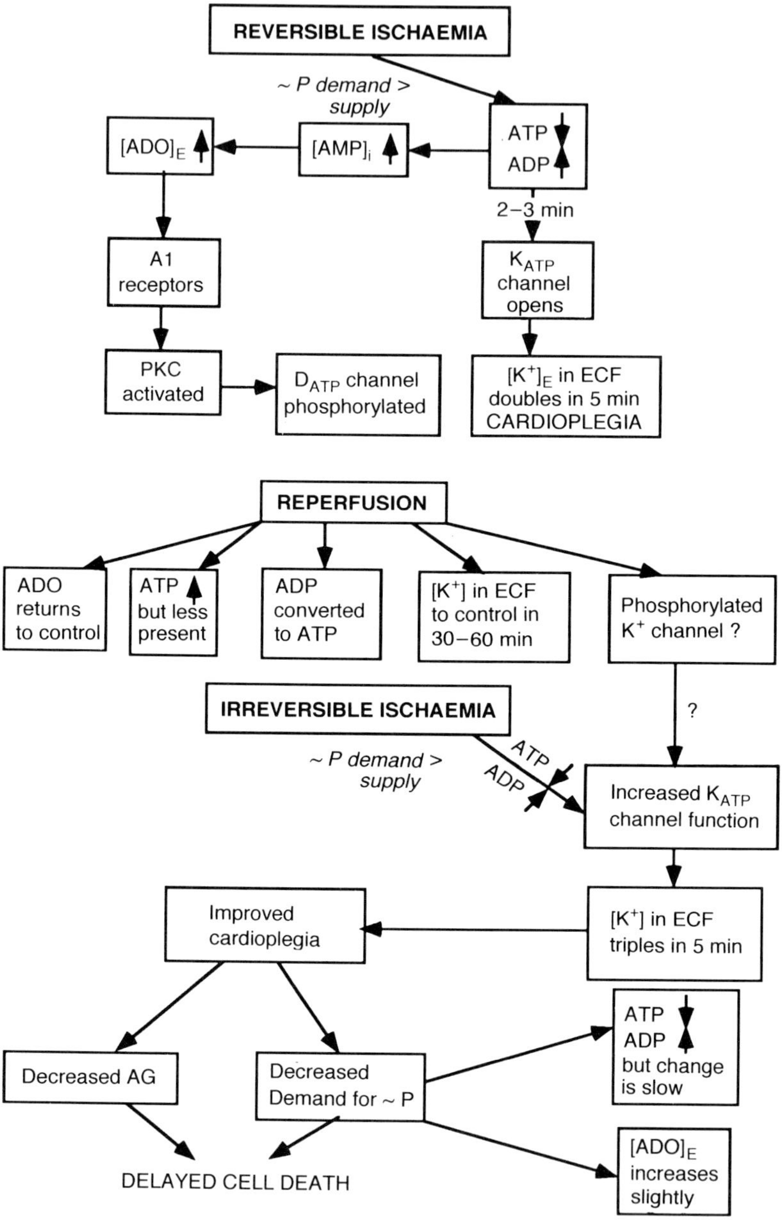

Figure 5. Diagram of a hypothetical sequence that might be involved in preconditioning with ischaemia in which the final mediator of the effect is proposed to be increased $[K]_E$ (extracellular K^+ in mmol K^+ l^{-1} ECF). Many of the steps in this sequence are known to occur, such as the rise in extracellular ADO ($[ADO]_E$), the rise in intracellular AMP ($[AMP]_i$) and ADP, the decrease in tissue ATP, and the increase in $[K^+]_E$. However, several critical steps are unknown. These include the question of activation of PKC in large animal hearts as well as

Thus, although unproven, there are much data available to indicate that K^+ channel activation and preconditioning are related events.

8. Summary

Ischaemic preconditioning against lethal cell injury is summarized in variable detail. The major biologic features of preconditioning are: (i) it can be induced by brief episodes of ischaemia and reperfusion; (ii) it delays but does not prevent ischaemic cell death; (iii) it is a transient effect that lasts 60–120 min but is lost after 180 min of reperfusion; (iv) however, there seems to be a SWOP between 1 and 3 days; and (v) it is associated with reduced demand for high energy phosphate during a sustained episode of ischaemia. Evidence that ischaemic preconditioning involves endogenous mediators including ADO is presented, as is evidence that opening of the K_{ATP} channel may be the final mediator of the preconditioning effect.

A molecular explanation of the pathogenesis of preconditioning must account for the biology of the phenomenon. At the very least, it must account for the transient nature of the effect, how this change results in the typical time course of the preconditioning, and how the change or changes results in reduced energy demand in the ischaemic tissue. Also, it should explain why preconditioning delays the transition to cell death in myocytes destined to die within a zone of ischaemia.

Preconditioning with ischaemia also has protective effects against arrhythmias, stunning and sympathetic denervation. Since there is no proof that the mechanism of protection is identical in each of the protective effects ascribed to preconditioning, this phenomenon always should be defined in terms of the endpoint used.

Acknowledgements

This work was supported in part by grants from NIH HL23138 and HL27416.

---

the effects of PKC on the K_{ATP} channel. It is not known whether the channel is phosphorylated by PKC and whether phosphorylation alters channel function. More $[K^+]_E$ could appear in the extracellular space during the second episode by a variety of means that are unrelated to phosphorylation. For example, if the channels opened earlier, or if more K_{ATP} channels opened, or if more K_{ATP} channels stayed open. Clearly, much remains to be learned about the box labelled 'increased K_{ATP} channel function'. Reduction in Ca^{2+} entry during ischaemia probably mediates the energy-sparing effect. AG, anaerobic glycolysis.

References

Alkulaifi AM, Pugsley WB and Yellon DM. (1993) The influence of the time period between preconditioning, ischaemia and prolonged ischaemia on myocardial protection. *Cardioscience,* **4,** 163–169.

Allison TB and Holsinger JW Jr. (1983) Myocardial metabolism and regional myocardial blood flow in the canine left ventricle following twenty min of circumflex artery occlusion and reperfusion. *J. Mol. Cell. Cardiol.* **15,** 151–161.

Auchampach JA and Gross GJ. (1992) Blockade of ATP-sensitive potassium channels prevents myocardial preconditioning in dogs. *Circ. Res.* **70,** 223–233.

Bolli R, Patel BS, Jeroudi MO, Lai EK and McCay PB. (1988) Demonstration of free radical generation in 'stunned' myocardium of intact dogs with the use of the spin trap alpha-phenyl *N*-teriary butyl nitrone. *J. Clin. Invest.* **82,** 476–485.

Bunch FT, Thornton J, Cohen MV and Downey JM. (1992) Adenosine is an endogenous protectant against stunning during repetitive ischemic episodes in the heart. *Am. Heart J.* **124,** 1440–1446.

Cave AC and Hearse DJ. (1992) Ischaemic preconditioning and contractile function: studies with normothermic and hypothermic global ischaemia. *J. Mol. Cell. Cardiol.* **24,** 1113–1123.

Chien GL, Tang Y and Van Winkle DM. (1995) Stereospecific blockade of myocardial ischemic preconditioning by Naloxone. *Circulation,* **92** (suppl. I), I-389.

Coker SJ, Parratt JR, Ledingham IJ and Zeitlin IJ. (1981) Thromboxane and prostacyclin release from ischaemic myocardium in relation to arrhythmias. *Nature,* **291,** 323–324.

Currie RW, Karmazyn M, Kloc M and Mailer K. (1988) Heat-shock response is associated with enhanced postischemic ventricular recovery. *Circ. Res.* **63,** 543–549.

Deutsch E, Berger M, Kussmaul WG, Hirshfeld JW Jr, Herrmann HC and Laskey WK. (1990) Adaptation to ischemia during precutaneous transluminal coronary angioplasty. Clinical, hemodynamic, and metabolic features. *Circulation,* **82,** 2044–2051.

Downey JM, Cohen MV, Ytrehus K and Liu Y. (1994) Cellular mechanisms in ischemic preconditioning: the role of adenosine and protein kinase C. *Ann. NY Acad. Sci.* **723,** 82–98.

Downey JM, Thornton JD, Van Winkle DM, Stanley AWH and Olsson RA. (1990) Protection against infarction afforded by preconditioning is mediated by A_1 adenosine receptors in rabbit heart. *Circulation,* **84,** 350–356.

Fleet WF, Johnson TA, Graebner CA and Gettes LS. (1985) Effect of serial brief ischemic episodes on extracellular K^+, pH, and activation in the pig. *Circulation,* **72,** 922–932.

Grover GJ, Newburger J, Sleph PG, Dzwonczyk S, Taylor SC, Ahmed SZ and Atwal KSW. (1991) Cardioprotective effects of the potassium channel opener Cromakalim: stereoselectivity and effects on myocardial adenine nucleotides. *J. Pharmacol. Exp. Ther.* **2257,** 156–162.

Grover GJ, Sleph PG and Dzwonczyk S. (1992) Role of myocardial ATP-sensitive potassium channels in mediating preconditioning in the dog heart and their possible interaction with adenosine A1-receptors. *Circulation,* **86,** 1310–1316.

Hide EJ, Piper J, Ney P, Thiememann C and Vane JR. (1995) Endothelium-1 reduces infarct size in the rabbit by activating ATP-sensitive potassium channels. *Circulation,* **92** (suppl. I), I-455.

Hill JL and Gettes LS. (1980) Effect of acute coronary artery occlusion on local myocardial extracellular K^+ activity in swine. *Circulation,* **61,** 768–778.

Jennings RB and Reimer KA. (1996) Part II. Discovery and early history of preconditioning. In: *Intrinsic Adaptive Mechanisms during Ischemia and Reperfusion: Preconditioning, Stunning, Hibernation* (eds GR Heyndrickx, S Vatner and W Wijns). Futura, Aalst, Belgium.

Jennings RB, Reimer KA and Steenbergen C Jr. (1986) Myocardial ischemia revisited. The osmolar load, membrane damage, and reperfusion (Editorial). *J. Mol. Cell. Cardiol.* **18,** 769–780.

Jennings RB and Steenbergen C Jr. (1985) Nucleotide metabolism and cellular damage in myocardial ischemia. *Ann. Rev. Physiol.* **47,** 727–749.

Kirsch GE, Codina J, Birnbaumer L and Brown AM. (1990) Coupling of ATP-sensitive K^+

channels to A_1 receptors by G proteins in rat ventricular myocytes. *Am. J. Physiol.* **259,** H820–H826.

Kitakaze M, Hori M, Takashima S, Sato H, Inoue M and Kamada T. (1993) Ischemic preconditioning increases adenosine release and 5′-nucleotidase activity during myocardial ischemia and reperfusion in dogs. Implications for myocardial salvage. *Circulation,* **87,** 208–215.

Lawson CS, Coltart DJ and Hearse DJ. (1993) The antiarrhythmic action of ischaemic preconditioning in rat hearts does not involve functional G_i proteins. *Cardiovasc. Res.* **27,** 681–687.

Lederer WJ and Nichols CG. (1995) Regulation and function of adenosine triphosphate-sensitive potassium channels in the cardiovascular system. In: *Cardiac Electrophysiology. From Cell to Bedside,* 2nd edn (eds DP Ziprd and J Jalife). WB Saunders, Philadelphia, PA, pp. 103–115.

Liu Y and Downey JM. (1992) Ischemic preconditioning protects against infarction in rat heart. *Am. J. Physiol.* **263,** H1107–H1112.

Liu GS, Thornton J, Van Winkle DM, Stanley AWH, Olsson RA and Downey JM. (1991) Protection against infarction afforded by preconditioning is mediated by A1 adenosine receptors in rabbit heart. *Circulation,* **84,** 350–356.

Liu Y, Ytrehus K, Downey JM. (1994) Evidence that translocation of protein kinase C is a key event during ischaemic preconditioning of rabbit myocardium. *J. Mol. Cell. Cardiol.* **26:** 661–668.

Liu Y, Tsuchida A, Cohen MV, Downey JM. (1995) Pretreatment with angiotensin II activates protein kinase C and limits myocardial infarction in isolated rabbit hearts. *J. Mol. Cell. Cardiol.* **27:** 883–892.

Liu Y, Gao WD, O'Rourke B, Marban E. (1996) Synergistic modulation of ATP-sensitive K^+ currents by protein kinase C and adenosine. Implications for ischaemic preconditioning. *Circ. Res.* **78:** 443–454.

Louis J-C, Magal E and Yavin E. (1986) Protein kinase C alterations in the fetal rat brain after global ischemia. *J. Biol. Chem.* **263,** 19 282–19 285.

Marber MS, Latchman DS, Walker JM and Yellon DM. (1993) Cardiac stress protein elevation 24 hours following brief ischemia or heat stress is associated with resistance to infarction. *Circulation,* **88,** 1264–1272.

Martorana PA, Kettenbach B, Briepohl G, Linz W and Scholkens BA. (1990) Reduction of infarct size by local angiotensin-converting enzyme inhibition is abolished by a bradykinin antagonist. *Eur. J. Pharmacol.* **182,** 395–396.

Miura T, Goto M, Urabe K, Endoh A, Shimamoto K and Iimura O. (1991) Does myocardial stunning contribute to infarct size limitation by ischemic preconditioning? *Circulation,* **84,** 2504–2512.

Miura T, Ogawa T, Iwamoto T, Shimamoto K and Iimura O. (1992) Dipyridamole potentiates the myocardial infarct size-limiting effect of ischemic preconditioning. *Circulation,* **86,** 979–985.

Miura T, Ogawa T, Iwamoto T, Tsuchida A and Iimura O. (1990) Infarct size limiting effect of preconditioning: its duration and 'dose–response' relationship. *Circulation,* **82,** III–271.

Miyazaki T and Zipes DP. (1989) Protection against autonomic denervation following acute myocardial infarction by preconditioning ischemia. *Circ. Res.* **64,** 437–448.

Mizumura T, Nithipatikom K and Gross GJ. (1995) Effects of nicroandil and glyceryl trinitrate on infarct size, adenosine release, and neutrophil infiltration in the dog. *Cardiovasc. Res.* **29,** 482–489.

Murry CE, Jennings RB and Reimer KA. (1986) Preconditioning with ischemia: a delay of lethal cell injury in ischemic myocardium. *Circulation,* **74,** 1124–1136.

Murry CE, Richard VJ, Jennings RB and Reimer KA. (1991) Myocardial protection is lost before contractile function recovers from ischaemic preconditioning. *Am. J. Physiol. (Heart Circ. Physiol.)* **260,** H796–H804.

Murry CE, Richard VJ, Reimer KA and Jennings RB. (1990) Ischemic preconditioning slows energy metabolism and delays ultrastructural damage during sustained ischaemia. *Circ. Res.* **66,** 913–931.

Noda K, Sasaguri M, Ideishi M, Ikeda M and Arakawa K. (1993) Role of locally formed bradykinin in the reduction of myocardial infarct size in dogs. *Circ. Res.* **27,** 334–340.

Przyklenk K, Sussman MA, Simkhovich BZ and Kloner RA. (1995) Does ischemic preconditioning trigger translocation of protein kinase C in the canine model? *Circulation,* **92,** 1546–1557.

Reimer KA, Murry CE, Yamasawa I, Hill ML and Jennings RB. (1986) Four brief periods of myocardial ischemia cause no cumulative ATP loss or necrosis. *Am. J. Physiol. (Heart Circ. Physiol.)* **251,** H1306–H1315.

Richardt G, Waas W, Kranzhofer R, Cheng B, Lohse MJ and Schomig A. (1989) Interaction between the release of adenosine and noradrenaline during sympathetic stimulation: a feed-back mechanism in rat heart. *J. Mol. Cell. Cardiol.* **21,** 269–277.

Sack S, Mohri M, Arras M, Schwarz ER and Schaper W. (1993) Ischaemic preconditioning-time course of renewal in the pig. *Cardiovasc. Res.* **27,** 551–555.

Schultz JEJ, Rose E, Yao Z and Gross GJ. (1995) Evidence for involvement of opioid receptors in preconditioning in rat hearts. *Am. J. Physiol. (Heart Circ. Physiol.)* **268,** H2157–H2161.

Schulz R, Rose J and Heusch G. (1994) Involvement of activation of ATP-dependent potassium channels in ischemic preconditioning in swine. *Am. J. Physiol. (Heart Circ. Physiol.)* **267,** H1341–H1352.

Schwartz LM, Jennings RB and Reimer KA. (1995) Minimum ischemic stress required for myocardial preconditioning in dogs; effect of analgesia, using the opiate agonist-antagonist, butorphanol. *FASEB J.* **9,** A299.

Schwarz ER, Fleischhauer J, Schutz T, Foresti M, Chakupurakal R, vom Dahl J and Klues HG. (1995) Angiotensin II receptor antagonism reduces infarct size in the pig heart. *Circulation,* **92,** I-454–I-455.

Steenbergen C Jr, Perlman ME, London RE and Murphy E. (1993) Mechanism of preconditioning. Ionic alterations. *Circ. Res.* **72,** 112–125.

Swain JL, Sabina RL, Hines JJ, Greenfield JC Jr and Holmes EW. (1984) Repetitive episodes of brief ischemia (12 min) do not produce a cumulative depletion of high energy phosphate compounds. *Cardiovasc. Res.* **18,** 264–269.

Thornton JD, Liu GS, Olsson RA and Downey JM. (1992) Intravenous pretreatment with A1-selective adenosine analogues protects the heart against infarction. *Circulation,* **85,** 659–665.

Thornton JD, Striplin S, Liu GS, Swafford A, Stanley WH, Van Winkle DM and Downey JM. (1990) Inhibition of protein synthesis does not block myocardial protection afforded by preconditioning. *Am. J. Physiol. (Heart Circ. Physiol.)* **259,** H1822–H1825.

Van Wylen DGL. (1994) Effect of ischemic preconditioning on interstitial purine metabolite and lactate accumulation during myocardial ischemia. *Circulation,* **89,** 2283–2289.

Vander Heide RS, Hill ML, Reimer KA and Jennings RB. (1996) Effect of reversible ischemia on the activity of the mitochondrial ATPase: relationship to ischemic preconditioning. *J. Mol. Cell. Cardiol.* **28,** 103–112.

Vegh A, Komori S, Szekeres L and Parratt JR. (1992) Antiarrhythmic effects of preconditioning in anaesthetised dogs and rats. *Cardiovasc. Res.* **26,** 487–495.

Vogt AM, Podzuweit T, Ando H, Elsasser A, Arras M, Htun P and Schaper W. (1995) Lack of myocardial adenosine production causes refractoriness after ischemic preconditioning's protection has waned. *Circulation,* **92,** I-524.

Williams DO, Boatwright RB, Rugh KSW, Garner HE and Griggs DM Jr. (1991) Myocardial blood flow, metabolism, and function with repeated brief coronary occlusions in conscious ponies. *Am. J. Physiol. (Heart Circ. Physiol.)* **260,** H100–H109.

Yang X-M, Arnoult S, Tsuchida A, Cope D, Thornton JD, Daly JF, Cohen MV and Downey JM. (1993) The protection of ischaemic preconditioning can be reinstated in the rabbit heart after the initial protection has waned. *Cardiovasc. Res.* **27,** 556–558.

Yellon DM and Latchman DS. (1992) Stress proteins and myocardial protection during ischemia and reperfusion. In: *Myocardial Protection: the Pathophysiology of Reperfusion and Reperfusion Injury* (eds DM Yellon and RB Jennings). Raven Press, New York, pp. 185–195.

Yellon DM, Pasini E, Cargnoni A, Marber MS, Latchman DS and Ferrari R. (1992) The protective role of heat stress in the ischaemic and reperfused rabbit myocardium. *J. Mol. Cell. Cardiol.* **24,** 895–907.

Ytrehus K, Liu Y and Downey JM. (1994) Preconditioning protects ischemic rabbit heart by protein kinase C activation. *Am. J. Physiol. (Heart Circ. Physiol.)* **266,** H1145–H1152.

2

Mechanisms of preconditioning: correlates and epiphenomena

James M. Downey and Michael V. Cohen
Department of Physiology and Medicine, University of South Alabama, Mobile, AL 36688, USA

1. Introduction

When the heart is exposed to a brief period of ischaemia it adapts to be resistant to infarction from a subsequent period of ischaemia. This phenomenon has been termed ischaemic preconditioning. There has been a concerted effort to identify the mechanism of preconditioning with the hope that an intervention can be developed which would confer this protection to the patient suffering from an acute myocardial infarction.

2. Preconditioning's putative signal transduction pathway

In 1991 we made the discovery that adenosine released from the ischaemic heart triggers preconditioning (Liu *et al.*, 1991). Blocking

Ischaemia: preconditioning and adaptation, edited by M.S. Marber and D.M. Yellon

adenosine receptors blocked the protection from ischaemic preconditioning in rabbit hearts. Similarly, transient exposure to adenosine or an adenosine analogue selective for the A_1 adenosine receptor put the heart into a preconditioned state. Adenosine's involvement has been demonstrated in preconditioning the pig (Schwarz *et al.*, 1991; Van Winkle and Davis, 1981), dog (Kitakaze *et al.*, 1991), human (Ikonomidis *et al.*, 1995; Walker *et al.*, 1985) but not the rat (Li and Kloner, 1993; Liu and Downey, 1992) heart. Follow-up studies indicated that protein kinase C (PKC) was also involved in this protection. Protection could be blocked in the small animal hearts by staurosporine and polymyxin B (Ytrehus *et al.*, 1994), chelerytherine (Speechly-Dick *et al.*, 1994), and calphostin C (Armstrong *et al.*, 1994). Similarly, activation of PKC by phorbol esters (Armstrong and Ganote, 1994; Ytrehus *et al.*, 1994) or by diacylglycerol (Mitchell *et al.*, 1995; Speechly-Dick *et al.*, 1994; Ytrehus *et al.*, 1994) was found to mimic preconditioning. In addition we have found that stimulation of all of the known PKC-coupled receptors in the heart was also capable of mimicking preconditioning. Although PKC's role in preconditioning may be controversial in the pig and dog hearts (Kitakaze *et al.*, 1994; Przyklenk *et al.*, 1995; Vogt *et al.*, 1994) several recent studies indicate that activation of PKC does seem to be required for preconditioning's protection in both human atria (Speechly-Dick *et al.*, 1995) and ventricular myocytes (Ikonomidis *et al.*, 1995).

A similar involvement has been found for the adenosine triphosphate (ATP)-sensitive (K_{ATP}) potassium channel. In species like the dog K_{ATP} channel blockers selectively block protection while K_{ATP} openers mimic preconditioning (Gross and Auchampach, 1992; Grover *et al.*, 1992a). Under certain anaesthetics the same can be seen in the rabbit (Walsh *et al.*, 1994) but not in the rat (Grover *et al.*, 1992b; Liu and Downey, 1992). Taken together, the above evidence indicates that adenosine released from the ischaemic heart puts the myocytes into a preconditioned state by occupying A_1 receptors. These may well act to open K_{ATP} channels by activation of PKC.

3. Unanswered questions

While we now have some insight into the signal transduction system involved in this protective adaptation, the overall mechanism remains enigmatic. If opening of the K_{ATP} channel is the final event then why would opening of these channels be so protective? Shortening of the action potential and thus reduced calcium entry is often offered as an explanation but we see protection from pinacidil in non-beating isolated myocytes which obviously have no action potential (Armstrong *et al.*, 1995). Equally important is, what constitutes the memory in ischaemic preconditioning? The window of protection lasts for about an hour after the preconditioning event and then the heart reverts back to its non-preconditioned state. Timing studies exclude the simple explanation that the memory corresponds to the time when the PKC's substrate is phosphorylated because blocking phosphorylation with staurosporine during the preconditioning ischaemia did not prevent the transition to the preconditioned state (Liu *et al.*, 1994a). We have speculated that the memory is due to the transient presence of PKC in the membrane (Liu *et al.*, 1994a) but attempts to corroborate that hypothesis have so far been negative (Przyklenk *et al.*, 1995).

4. What is different in preconditioned myocardium: correlates and epiphenomena

A key step towards understanding the preconditioning mechanism would be to answer the simple question: what is the difference between a preconditioned and a non-preconditioned heart. The difficulty with answering this question is in separating epiphenomena (things that have changed but are unrelated to the protection) from the actual factors that are involved.

For example an ischaemic preconditioning protocol usually stuns the heart. Because stunned myocardium does less physical work it is reasonable to speculate that stunning might cause protection by reducing metabolic requirements during ischaemia. It was easy to show that stunning is only an epiphenomenon since not all preconditioned myocardium was stunned (Miura *et al.*, 1991) and under some conditions myocardium that was stunned was not preconditioned (Murry *et al.*, 1991).

A similar case can be made for glycogen depletion. Cardiac glycogen is depleted following a transient period of ischaemia. It has recently been hypothesized that glycogen depletion could be related to the mechanism of protection (Wolfe *et al.*, 1993). This is reasonable since preconditioned hearts are known to have a reduced rate of glycolysis (Murry *et al.*, 1990). However, a cause-and-effect relationship between glycogen content and protection has not been demonstrated. For example, is glycogen also depleted when the heart is pharmacologically preconditioned with an agent like adenosine or a K_{ATP} opener? Conversely all interventions which deplete glycogen do not seem to precondition (Lagerstrom *et al.*, 1988). Thus glycogen depletion may well turn out to be an epiphenomenon.

Recently Kolacassides *et al.* (1995) reported that preconditioned rat hearts go into contracture quicker during ischaemia than non-preconditioned hearts. Is early contracture a characteristic of all preconditioned hearts or does this simply reflect a residual energy deficit from the preconditioning ischaemia? Adenosine A_1 receptor agonists, which reportedly can precondition the rat heart against infarction (Liu and Downey, 1992), delay rather than hasten the onset of ischaemic contracture in that species (Lasley *et al.*, 1994). Thus early contracture may again be an epiphenomenon unrelated to the actual mechanism of preconditioning's protection.

5. Preconditioning and adenosine release patterns

A similar situation is seen with the amount of adenosine that the heart releases during ischaemia. While one group reports that adenosine release is augmented during ischaemia in preconditioned dog heart (Kitakaze *et al.*, 1993), others actually saw a decreased adenosine release during ischaemia if the heart had first been preconditioned (Bradamante *et al.*, 1995; Murry *et al.*, 1990; Van Wylen, 1994). Much of the adenosine released from the heart during ischaemia is thought to derive from ATP degradation. If a reduced rate of ATP degradation during ischaemia was responsible for the protection then this should be reflected in a reduced adenosine release. We therefore undertook a study to examine release patterns in the isolated rabbit heart.

To approach this question we measured purine release from isolated rabbit hearts exposed to 5 min of global ischaemia (Goto *et al.*, 1996), an intervention known to put these hearts into a preconditioned state (Liu *et al.*, 1994b). The effluent was collected in the first few minutes of reperfusion following each occlusion and analysed for adenosine and its breakdown products, inosine, hypoxanthine and xanthine. No xanthine was detected, which is not surprising since rabbit heart lacks xanthine oxidase (Downey *et al.*, 1988). To account for all of the released adenosine we added the amounts of adenosine, inosine and hypoxanthine in each sample which will be referred to as total purine. *Figure 1* reveals that

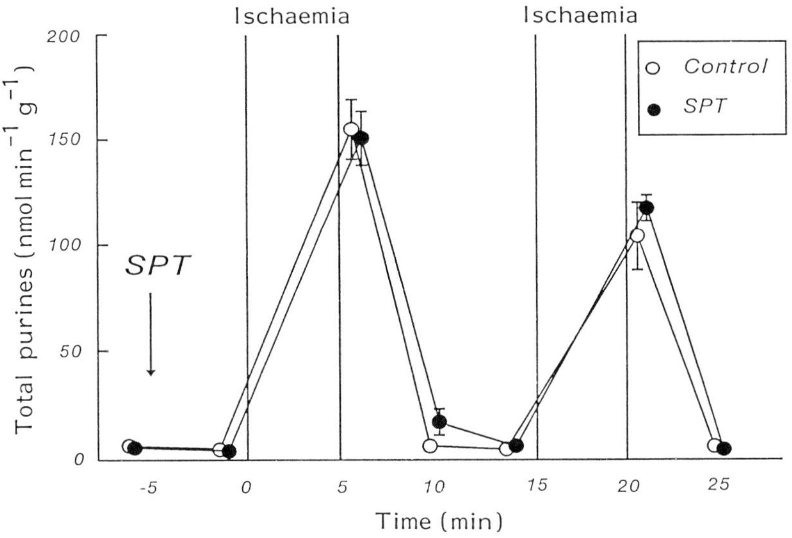

Figure 1. Total purine release rates following two serial 5 min occlusions in isolated rabbit hearts. Note that more adenosine is released following the first occlusion than by the second in the control hearts. Blocking preconditioning's protection does not prevent the attenuation of purine release following the second occlusion. Reproduced with permission from Goto *et al.* (1996) Attenuated purine production during subsequent ischaemia in preconditioned rabbit myocardium is unrelated to the mechanism of protection. *J. Mol. Cell Cardiol.* **28,** 447–454.

about 50% more purine was released following the first 5 min occlusion period than after the second occlusion period. Thus the preconditioned rabbit heart clearly releases less purine during ischaemia than the non-preconditioned rabbit heart. Note, however, that blocking preconditioning's protection with the adenosine receptor blocker 8-*p*-sulphophenyl adenosine does not prevent the attenuation of purine release with a second coronary occlusion. The presence of attenuated purine release in a non-protected heart suggests that the decreased release was unrelated to the protection.

Figure 2 shows that pharmacologically preconditioning the heart with a transient period of either adenosine (Liu *et al.*, 1994b) or phenylephrine (Tsuchida *et al.*, 1994) exposure also does not modify the purine release pattern. If the purine release was related to the mechanism of protection then a reduced amount of purine would be expected to be released on the first occlusion with a similar amount following the second occlusion, since the heart would have been in a preconditioned state for both

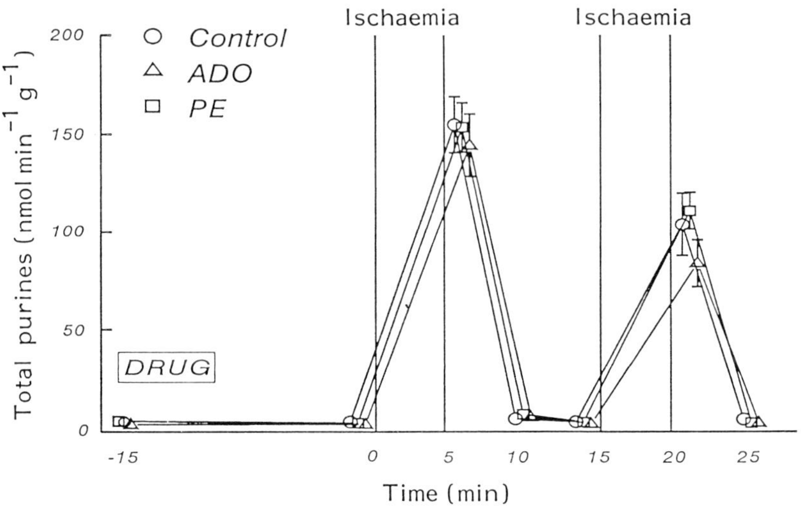

Figure 2. Total purine release rate following two serial 5 min coronary occlusions in isolated rabbit hearts which have been pharmacologically preconditioned with either adenosine (ADO) or phenylephrine (PE). Despite the fact that the first occlusion occurred while the hearts were preconditioned after the first occlusion they still released a similar amount of purines as a non-preconditioned heart (see *Figure 1*). They also released much less following the second occlusion. Reproduced with permission from Goto *et al.* (1996) Attenuated purine production during subsequent ischaemia in preconditioned rabbit myocardium is unrelated to the mechanism of protection. *J. Mol. Cell Cardiol.* **28,** 447–454.

occlusions. Note, however, that we again saw a large release of purine followed by an attenuated release. The heart simply seems to release less purine on a second period of ischaemia regardless of the state of preconditioning. We, therefore, must relegate the attenuation of purine release to the realm of an epiphenomenon.

6. Preconditioning and ST-segment evolution in pig heart

Several years ago it was reported that patients undergoing percutaneous transluminal angioplasty (PCTA) showed much less ST-segment shift in their electrocardiograms on subsequent occlusions of the coronary artery than on the first occlusion (Deutsch *et al.,* 1990). Cribier and colleagues challenged this notion and pointed out that the collateral flow was also often increased on subsequent coronary occlusions in PCTA patients and that the improved perfusion could have accounted for the improvement (Cribier *et al.*, 1992). If a reduced ST-segment is a marker for a transition to the preconditioned state then recordings from PCTA patients could be a valuable tool for assessing interventions which are proposed to precondition the human heart. We therefore asked: does preconditioning an animal's heart also reduce the ST-voltage during ischaemia?

We assessed the ST-segment patterns in open chest pigs (Shattock *et al.*, 1995) and rabbits (Cohen *et al.*, 1996). Both were chosen because they have a very sparse pattern of collateralization in their hearts. Epicardial electrodes were attached to the hearts in the ischaemic zone. Coronary collateral flow was measured during each ischaemic period in the pigs by measuring the washout rate of ^{133}Xe gas. In the pigs we occluded the coronary artery three times. The first two occlusions consisted of 8 min of coronary occlusion each followed by 8 min of reperfusion. The hearts then were subjected to 60 min of occlusion followed by 2 h of reperfusion to induce an infarct. The ST-segment voltage was zero prior to the first coronary occlusion. With the onset of occlusion the voltage from the epicardial electrode increased with time such that by 8 min it had reached almost 10 mV. *Figure 3* reveals that there were no differences in the voltages between the first, second or third occlusions after 8 min of regional ischaemia. However, note that the ST-segment voltage began to rise later in the second and, especially, the third occlusions such that in

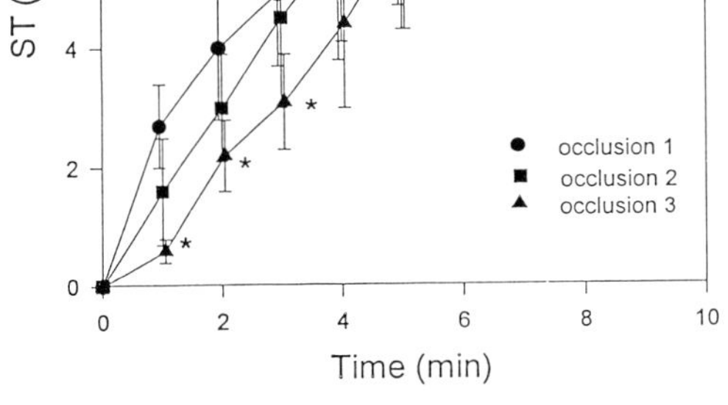

Figure 3. The evolution of the ST-segment from three serial 8 min coronary occlusions in open-chest pigs. Epicardial electrograms were used. Note a much smaller ST-segment voltage in the first 3 min of occlusion during the second and third occlusions.

the first 3 min significant difference existed between the third and the first occlusion.

The xenon washout rates revealed that collateral flow was low in all of the pigs and did not differ between the three occlusion periods. Furthermore, measurement of infarct size revealed that these pigs were preconditioned by the 2 × 8 min occlusion protocol. In the control hearts 42 ± 3% of the risk region was infarcted while only 14 ± 9% was infarcted in the preconditioned hearts. These data do support the concept that, at least early in the occlusion, the ST-segment voltage is reduced in preconditioned hearts. Remember that in PCTA patients coronary occlusions rarely exceed 2 min so the pig data are completely compatible with the human data.

7. Preconditioning and ST-segment evolution in rabbit heart

We next extended these studies to rabbits where we could pharmacologically manipulate the preconditioning process more easily (Cohen *et al.*, 1996). The preparation was virtually identical to that in the

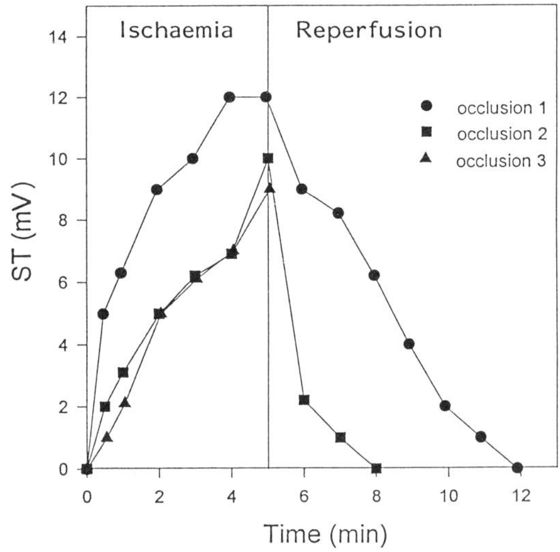

Figure 4. The evolution of the ST-segment from three serial 5 min coronary occlusions in an open-chest rabbit. Epicardial electrograms were used. Note a much smaller ST-segment voltage occurred at each time point during the second and third occlusions. This pattern was typical for all of the animals.

pig except that collateral flow was not measured. Our experience has been that collateral flow is very low in rabbits and does not change with serial occlusions (Cohen *et al.*, 1994). The protocol consisted of two 5 min occlusions each separated by 10 min of reperfusion. The third occlusion was for 30 min and hearts were reperfused for 3 h. *Figure 4* shows a typical experiment. ST-segment voltage rose with time with each occlusion but with the second and third occlusions the rise was much less. These voltages were quantitated by calculating the area under the voltage vs. time curve. The second and third occlusions had a significantly smaller area than the first (29.5 ± 5.0 mV . min and 28.2 ± 4.7 mV . min vs. 45.8 ± 5.4 mV . min, respectively). When the protection of ischaemic preconditioning was blocked by intravenous infusion of 8-(*p*-sulphophenyl)theophylline, an adenosine receptor antagonist, the attenuation of ST-segment elevation previously observed after the first occlusion was no longer apparent (42.8 ± 6.7, 35.7 ± 4.0 and 37.7 ± 6.0 mV . min, for the three occlusions). A large infarct size in those hearts revealed that protection had been blocked.

When preconditioning was triggered by locally releasing cardiac catecholamines by tyramine injection (Thornton *et al.*, 1993a) rather than ischaemia, there also was no significant alteration in ST-segment elevation areas during the three occlusion periods (26.2 ± 3.2, 24.8 ± 2.9, and 24.2 ± 2.1 mV . min, respectively). Therefore, when the myocardium can be preconditioned by ischaemia, the ST-elevation is diminished during subsequent episodes of ischaemia. But if preconditioning was prevented pharmacologically, then the attenuation of ST-elevation was also abolished. When the myocardium was pharmacologically preconditioned, ST-segment elevation during the first occlusion in the previously preconditioned heart was already attenuated so that no change in ST-elevation was observed during subsequent occlusions. Thus the attenuation in ST-voltage is not an epiphenomenon but rather seems to be intimately associated with the presence or absence of a protected state.

It is interesting to note that *Figure 4* shows a more rapid recovery of the ST-segment during the second reperfusion. That pattern persisted in all of the above groups, indicating that a more rapid recovery of the ST-segment during reperfusion is only an epiphenomenon and occurs on the second occlusion regardless of the state of protection.

The ST-segment is produced by a loss of phase 4 membrane potential in the ischaemic region. These data would suggest that the membrane potential is lost more slowly during ischaemia if the heart is in a preconditioned state and that this delay is central to the protective mechanism. Preconditioning has been linked to opening of the K_{ATP} channel (Gross and Auchampach, 1992) which is certainly a major determinant of membrane potential but how that opening could result in preservation of a normal membrane potential during ischaemia is not obvious. During ischaemia the membrane should be very near its potassium equilibrium potential and the progressive depolarization should reflect the accumulation of sodium.

8. Adenosine fails to protect the virgin heart

Let us dig deeper into the question as to what is different in the preconditioned heart. From a functional standpoint we would say that: in the preconditioned heart a coupling has been established between adenosine receptors and the end effector. Adenosine acting as an agonist of the A_1 receptor was the first autacoid to be proposed as a trigger for the anti-infarct effect of preconditioning. Adenosine is released in quantity during ischaemia so much so that it achieves micromolar levels within a minute or two of occlusion. This quickly saturates the A_1 receptors whose K_d is in the nanomolar range. Yet the virgin heart is not protected by this adenosine. Even when the adenosine concentration during ischaemia was augmented 100-fold by a deaminase inhibitor, adenosine still failed to protect the non-preconditioned heart (Silva *et al.*, 1995). Furthermore, adenosine receptor blockers have no effect on infarct size in the non-preconditioned rabbit heart (Liu *et al.*, 1991) again showing that endogenous adenosine simply does not protect against necrosis in the virgin heart. If the heart has been preconditioned, however, then the adenosine that is released becomes very protective. Blocking adenosine receptors after preconditioning but just prior to the long ischaemia, completely aborts the protection (Thornton *et al.*, 1993b) indicating that adenosine receptors not only serve as a trigger for the onset of the preconditioned state but also must be reactivated during the long ischaemia in order to protect.

9. PKC fails to protect the virgin heart

We now believe that adenosine receptors protect by activating PKC in the myocyte. PKC inhibitors block the protective effect of ischaemic preconditioning, and agonists of any of the PKC-coupled receptors in the heart (adrenergic α_1, angiotensin II AT_1, endothelin ET_1, and adenosine A_1) can fully mimic preconditioning. The timing of when PKC must act is also interesting. Blocking PKC in the non-preconditioned rabbit heart has no effect on infarct size, as was also seen with adenosine blockers. Bracketing the preconditioning ischaemia with the PKC blocker staurosporine also has no effect against protection, but when the long ischaemia occurs in the presence of staurosporine, protection is completely blocked. The parallel to what is seen with the adenosine receptor blockers seems unmistakable. If the preconditioned heart is one in which a coupling has been established between the adenosine receptor and the end effector, and if PKC is an intermediate in that pathway, then the question arises; is that coupling established upstream or downstream of PKC? The data are not clear on this point. Whatever the case, it seems to be the newly established coupling that makes the heart's endogenous adenosine become protective.

10. Phenylephrine can establish the adenosine–PKC coupling

The coupling theory is strongly supported by observations with the α_1 agonist, phenylephrine. Transient exposure to phenylephrine fully preconditions the heart but this protection can be blocked if an adenosine blocker is given just prior to the long ischaemia. Blocking adenosine receptors only during the phenylephrine exposure does not block protection. Thus, transient activation of α_1 receptors also makes the heart become sensitive to its own adenosine. It could be illustrated that the coupling was to PKC by adding phenylephrine along with the adenosine antagonist during the long ischaemia in the above experiment. In this situation protection returned, presumably because the phenylephrine was able to activate PKC directly despite the adenosine block.

11. Does the translocation theory explain the data?

The mechanism for the establishment of the coupling between adenosine receptors and PKC is unknown. We have proposed that this coupling may be a result of translocation of a critical PKC isoform into the membranes where it can be quickly activated. The attractiveness of the translocation theory is that translocation, the first step in activation, does not involve phosphorylation and can occur in the presence of staurosporine. Experimental evidence for or against this theory is currently absent, however. We believe that coupling represents a critical part of the preconditioning puzzle and should receive more attention in future studies.

References

Armstrong S and Ganote CE. (1994) Preconditioning of isolated rabbit cardiomyocytes: effects of glycolytic blockade, phorbol esters, and ischaemia. *Cardiovasc. Res.* **28,** 1700–1706.

Armstrong S, Downey JM and Ganote CE. (1994) Preconditioning of isolated rabbit cardiomyocytes: induction by metabolic stress and blockade by the adenosine antagonist SPT and calphostin C, a protein kinase C inhibitor. *Cardiovasc. Res.* **28,** 72–77.

Armstrong SC, Liu GS, Downey JM and Ganote CE. (1995) Potassium channels and preconditioning of isolated rabbit cardiomyocytes: effects of glyburide and pinacidil. *J. Mol. Cell Cardiol.* **27,** 1765–1774.

Bradamante S, Piccinini F, Delu C, Janssen M and de Jong JW. (1995) NMR evaluation of changes in myocardial high energy metabolism produced by repeated short periods of ischemia. *Biochim. Biophys. Acta* **1243,** 1–8.

Cohen MV, Yang X-M, Liu Y, Snell KS and Downey JM. (1994) A new animal model of controlled coronary artery occlusion in conscious rabbits. *Cardiovasc. Res.* **28,** 61–65.

Cohen MV, Yang X-M and Downey JM. (1996) Attenuation of ST-segment elevation during repetitive coronary occlusions reflects the protection of ischemic preconditioning. *Cardiovasc. Res.*, in press.

Cribier A, Korsatz L, Koning R, Rath P, Gamra H, Stix G, Merchant S, Chan C and Letac B. (1992) Improved myocardial ischemic response and enhanced collateral circulation with long repetitive coronary occlusion during angioplasty: a prospective study. *J. Am. Coll. Cardiol.* **20,** 578–586.

Deutsch E, Berger M, Kussmaul WG, Hirshfeld JW, Jr, Herrmann HC and Laskey WK. (1990) Adaptation to ischemia during percutaneous transluminal coronary angioplasty: clinical, hemodynamic, and metabolic features. *Circulation* **82,** 2044–2051.

Downey JM, Hearse DJ and Yellon DM. (1988) The role of xanthine oxidase during myocardial ischemia in several species including man. *J. Mol. Cell Cardiol.* **20** (Suppl. II), 55–63.

Goto M, Cohen MV, Van Wylen DGL and Downey JM. (1996) Attenuated purine production during subsequent ischemia in preconditioned rabbit myocardium is unrelated to the mechanism of protection. *J. Mol. Cell Cardiol.*, **28,** 447–454.

Gross GJ and Auchampach JA. (1992) Blockade of ATP-sensitive potassium channels prevents myocardial preconditioning in dogs. *Circ. Res.* **70,** 223–233.

Grover GJ, Sleph PG and Dzwonczyk S. (1992a) Role of myocardial ATP-sensitive potassium channels in mediating preconditioning in the dog heart and their possible interaction with adenosine A1-receptors. *Circulation*, **86,** 1310–1316.

Grover GJ, Dzwonczyk S and Sleph PG. (1992b) ATP-sensitive K^+ channel activation does not mediate preconditioning in isolated rat hearts (abstract). *Circulation,* **86** (Suppl. I), I–341.

Ikonomidis JS, Shirai T, Weisel RD, Rao V, Mickle DAG and Li R-K. (1995) Signal transduction in human cardiomyocyte preconditioning (abstract). *J. Mol. Cell Cardiol.* **27,** A40.

Kitakaze M, Hori M, Takashima S, Sato H and Kamada T. (1991) Augmentation of adenosine production during ischemia as a possible mechanism of myocardial protection in ischaemic preconditioning (abstract). *Circulation,* **84** (Suppl. II), II–306.

Kitakaze M, Hori M, Takashima S, Sato H, Inoue M and Kamada T. (1993) Ischemic preconditioning increases adenosine release and 5′-nucleotidase activity during myocardial ischaemia and reperfusion in dogs: implications for myocardial salvage. *Circulation,* **87,** 208–215.

Kitakaze M, Minamino T, Shinozaki Y, Sakamoto H, Mori H, Kurihara T and Hori M. (1994) Activation of protein kinase C and subsequent activation of ectosolic 5′-nucleotidase as a major cause for the infarct size-limiting effect of ischemic preconditioning (abstract). *Circulation,* **90** (Suppl. I), I–207.

Kolocassides KG, Galinanes M and Hearse DJ. (1995) Preconditioning accelerates contracture and ATP depletion in blood perfused rat hearts. *Am. J. Physiol.* **269,** H1415–H1420.

Lagerstrom CF, Walker WE and Taegtmeyer H. (1988) Failure of glycogen depletion to improve left ventricular function of the rabbit heart after hypothermic ischemic arrest. *Circ. Res.* **63,** 81–86.

Lasley RD, Noble MA, Paulsen KL and Mentzer JMR. (1994) Adenosine attenuates phorbol ester-induced negative inotropic and vasoconstrictive effects in rat hearts. *Am. J. Physiol.* **266,** H2159–H2166.

Li Y and Kloner RA. (1993) The cardioprotective effects of ischemic "preconditioning" are not mediated by adenosine receptors in rat hearts. *Circulation,* **87,** 1642–1648.

Liu Y and Downey JM. (1992) Ischemic preconditioning protects against infarction in rat heart. *Am. J. Physiol.* **263,** H1107–H1112

Liu GS, Thornton J, Van Winkle DM, Stanley AWH, Olsson RA and Downey JM. (1991) Protection against infarction afforded by preconditioning is mediated by A1 adenosine receptors in rabbit heart. *Circulation,* **84,** 350–356.

Liu Y, Ytrehus K and Downey JM. (1994a) Evidence that translocation of protein kinase C is a key event during ischemic preconditioning of rabbit myocardium. *J. Mol. Cell Cardiol.* **26,** 661–668.

Liu GS, Richards SC, Olsson RA, Mullane K, Walsh RS and Downey JM. (1994b) Evidence that the adenosine A3 receptor may mediate the protection afforded by preconditioning in the isolated rabbit heart. *Cardiovasc. Res.* **28,** 1057–1061.

Mitchell MB, Meng X, Ao L, Brown JM, Harken AH and Banerjee A. (1995) Preconditioning of isolated rat heart is mediated by protein kinase C. *Circ. Res.* **76,** 73–81.

Miura T, Goto M, Urabe K, Endoh A, Shimamoto K and Iimura O. (1991) Does myocardial stunning contribute to infarct size limitation by ischemic preconditioning? *Circulation,* **84,** 2504–2512.

Murry CE, Richard VJ, Reimer KA and Jennings RB. (1990) Ischemic preconditioning slows energy metabolism and delays ultrastructural damage during a sustained ischemic episode. *Circ. Res.* **66,** 913–931.

Murry CE, Richard VJ, Jennings RB and Reimer KA. (1991) Myocardial protection is lost before contractile function recovers from ischemic preconditioning. *Am. J. Physiol.* **260,** H796–H804.

Przyklenk K, Sussman MA, Simkhovich BZ and Kloner RA. (1995) Does ischemic preconditioning trigger translocation of protein kinase C in the canine model? *Circulation,* **92,** 1546–1557.

Schwarz ER, Mohri M, Sack S and Arras M. (1991) The role of adenosine and its A1-receptor in ischemic preconditioning (abstract). *Circulation,* **84** (Suppl. II), II–191.

Shattock MJ, Lawson CS, Hearse DJ and Downey JM. (1995) Do temporal electrophysiological and collateral flow changes mirror preconditioning-induced infarct size limitation in the pig? (abstract). *J. Mol. Cell Cardiol.* **27,** A143.

Silva PH, Dillon D and Van Wylen DGL. (1995) Adenosine deaminase inhibition augments interstitial adenosine but does not attenuate myocardial infarction. *Cardiovasc. Res.* **29,** 616–623.

Speechly-Dick ME, Mocanu MM and Yellon DM. (1994) Protein kinase C: its role in ischemic preconditioning in the rat. *Circ. Res.* **75,** 586–590.

Speechly-Dick ME, Grover GJ and Yellon DM. (1995) Does ischemic preconditioning in the human involve protein kinase C and the ATP-dependent K^+ channel? *Circ. Res.* **77,** 1030–1035.

Thornton JD, Daly JF, Cohen MV, Yang X-M and Downey JM. (1993a) Catecholamines can induce adenosine receptor-mediated protection of the myocardium but do not participate in ischemic preconditioning in the rabbit. *Circ. Res.* **73,** 649–655.

Thornton JD, Thornton CS and Downey JM. (1993b) Effect of adenosine receptor blockade: preventing protective preconditioning depends on time of initiation. *Am. J. Physiol.* **265,** H504–H508.

Tsuchida A, Liu Y, Liu GS, Cohen MV and Downey JM. (1994) Alpha1-adrenergic agonists precondition rabbit ischemic myocardium independent of adenosine by direct activation of protein kinase C. *Circ. Res.* **75,** 576–585.

Van Winkle DM and Davis RF. (1991) Ischemic preconditioning of myocardium: effect of the adenosine agonist phenylisopropyl adenosine (PIA)(abstract). *Circulation,* **84** (Suppl. II), II–305.

Van Wylen DGL. (1994) Effect of ischemic preconditioning on interstitial purine metabolite and lactate accumulation during myocardial ischemia. *Circulation,* **89,** 2283–2289.

Vogt A, Barancik M, Weihrauch D, Arras M, Podzuweit T and Schaper W. (1994) Activation of protein kinase C fails to protect ischemic porcine myocardium from infarction *in vivo* (abstract). *J. Mol. Cell Cardiol.* **26,** CXVIII.

Walker DM, Walker JM, Pugsley WB, Pattison CW and Yellon DM. (1995) Preconditioning in isolated superfused human muscle. *J. Mol. Cell Cardiol.* **27,** 1349–1357.

Walsh RS, Tsuchida A, Daly JJF, Thornton JD, Cohen MV and Downey JM. (1994) Ketamine-xylazine anaesthesia permits a KATP channel antagonist to attenuate preconditioning in rabbit myocardium. *Cardiovasc. Res.* **28,** 1337–1341.

Wolfe CL, Sievers RE, Visseren FLJ and Donnelly TJ. (1993) Loss of myocardial protection after preconditioning correlates with the time course of glycogen recovery within the preconditioned segment. *Circulation,* **87,** 881–892.

Ytrehus K, Liu Y and Downey JM. (1994) Preconditioning protects ischemic rabbit heart by protein kinase C activation. *Am. J. Physiol.* **266,** H1145–H1152.

3

Role of the ATP-sensitive potassium channel in ischaemic preconditioning

Gary J. Grover
Department of Cardiovascular Biochemistry, Bristol-Myers Squibb Pharmaceutical Research Institute, P.O. Box 4000, Princeton, NJ 08543-4000, USA

1. Introduction

The discovery of preconditioning in the myocardium has created much excitement among researchers in the field of myocardial ischaemia and reperfusion injury. The profound cardioprotective efficacy of preconditioning has led to a search for its mechanism, with the hope of developing novel therapeutics with the advantage of simulating an endogenous protective mechanism. While our understanding of the nature of preconditioning has improved, there is still much to be learned. One interesting school of thought on the mechanism of preconditioning concerns the role of adenosine triphosphate (ATP)-sensitive potassium channels (K_{ATP}). Garrett Gross' laboratory was the first to establish a role for K_{ATP} in preconditioning (Gross and Auchampach, 1992). The publication of their results was timely given the growing interest in the role of K_{ATP} in the pathogenesis of myocardial ischaemia. Work begun

Ischaemia: preconditioning and adaptation, edited by M.S. Marber and D.M. Yellon

more than 3 years prior to Gross's publication showed that K_{ATP} openers protected ischaemic myocardium, therefore these data provided an important bridge between preconditioning and the pharmacological work with K_{ATP} openers.

In this chapter, we describe the profile of myocardial preconditioning and use this as a basis for comparison with the profile of cardioprotection for K_{ATP} openers. If K_{ATP} opening is involved in the mechanism of this, then a pharmacological activator of this channel should simulate preconditioning and a blocker should abolish preconditioning. We will first describe data suggesting that the profile of activity of K_{ATP} openers is consistent with the profile of preconditioning. While such data suggest the possibility that K_{ATP} opening can be a mechanism for preconditioning, the determination of the effect of K_{ATP} blockers on preconditioning is necessary to prove such an involvement. The overall goal of this chapter is to give the reader information about the interaction of the K_{ATP} channel and preconditioning as well as to describe our current thinking on the function of K_{ATP} in low blood flow states.

2. Summary of the profile of myocardial preconditioning

Preconditioning is described as the protection observed after a prolonged ischaemic episode which is preceded by one or more short periods of ischaemia or a similar stress such as hypoxia. The protective effect of preconditioning is independent of haemodynamic alterations and appears to be an effect directly involving the myocardium (see review of Jennings et al., 1991). Studies have shown that increases in collateral blood flow do not explain the cardioprotective effects of preconditioning (Jennings et al., 1991). The protective effect of preconditioning is also not dependent on reductions in cardiac function (cardioplegic effect). The effect of preconditioning on myocardial energetics is still not clearly defined, although some studies have been done. A study by Murry *et al.* (1990) has shown that a short period of coronary occlusion in dogs results in some ATP depletion before the final, prolonged ischaemia. ATP during the longer ischaemia, however, falls at a slower rate in preconditioned animals. Early into the prolonged ischaemia, ATP in preconditioned dogs was higher, but this effect was slight and was not maintained. Therefore, the contribution that this slight ATP conservation has to overall protection is presently unclear. In addition, recent reports from another laboratory suggest that preconditioning can be observed despite a lack of preservation of ATP levels (Chen *et al.*, 1995)

3. Pharmacology of K_{ATP} openers in models of myocardial ischaemia

3.1. K_{ATP}: general considerations

K_{ATP} are metabolically regulated potassium channels which were first described in cardiac myocytes by Noma (1983). They are so named because ATP is one of the major gating mediators for this channel. ATP binds to a regulatory site on K_{ATP} and inhibits channel opening (see review by Edwards and Weston, 1993). When ATP levels are reduced, as might be seen during ischaemia, the probability of K_{ATP} channels opening increases. Numerous other factors which are also associated with metabolism are known to gate K_{ATP} and include: adenosine, adenosine diphosphate (ADP), arachidonic acid and pH (Edwards and Weston, 1993). K_{ATP} were first described in cardiac myocytes, but they have since been found in smooth muscle, skeletal muscle, kidney and various regions of the brain (Spruce *et al.*, 1985; Treherne and Ashford, 1991; de Weille *et al.*, 1988). While K_{ATP} appears to be similar in these tissues, it is possible that some differences occur. These may be helpful for developing pharmacological modulators of this channel which are tissue selective, and this will be discussed in more detail below.

Currently, much work is being done to elucidate the mechanism of regulation of K_{ATP} and a clearer picture is emerging (see review by Edwards and Weston, 1993). As stated earlier, an ATP regulatory or inhibitor site has been hypothesized in which ATP blocks K_{ATP} and this is not dependent on phosphorylation. Non-hydrolysable analogues of ATP are equally capable of inhibiting K_{ATP}. ADP reduces the sensitivity of the channel to ATP and it has been hypothesized that the ratio of ADP/ATP is important in determining channel opening. In the absence of ATP, however, ADP does inhibit K_{ATP} opening. A phosphorylation site is also hypothesized to exist on K_{ATP} as ATP is necessary to prevent channel run-down, a process thought to be due to channel dephosphorylation. It is known that Mg-ATP increases the opening probability of K_{ATP}. The regulation of K_{ATP} phosphorylation is presently unclear, but it is thought that cAMP-dependent protein kinase is involved in the signal transduction pathway (Light *et al.*, 1995; Ribalet *et al.*, 1989). The role for protein kinase C (PKC) is not as clear, but it is thought to have a modulatory role in gating K_{ATP} (de Weille *et al.* 1989).

Further work on the regulation of K_{ATP} suggested that this channel can be modulated by receptor–G protein interactions, possibly through purinergic receptors (Kirsch *et al.*, 1990). Recent work has identified proteins which act as ATP transporters belonging to the ATP binding cassette family of proteins (Aguilar-Bryan *et al.*, 1995; Al-Aqati, 1995; Philipsone and Steiner, 1995). Work with the cystic fibrosis transmembrane conductance regulator (CFTR) showed that this protein regulates a specific chloride channel, an effect which is blocked by the K_{ATP} blocker glyburide (see review by Al-Aqati, 1995). This ATP transporter releases ATP, which interacts with a purinergic receptor subtype which then opens chloride channels. A similar ATP permeant protein has been termed the sulphonylurea receptor (SUR), which is thought to have a similar function to CFTR, but is linked with K_{ATP} (Al-Aqati, 1995). It has been hypothesized that in smooth muscle cells, and perhaps insulin-secreting cells, K_{ATP} is tonically opened by ATP (or another factor) secreted by SUR and this occurs through a purinergic receptor. It has been shown by Kirsch *et al.* (1990) that adenosine can open K_{ATP} in neonatal rat cardiomyocytes, further indicating an interaction of K_{ATP} with purinergic receptors. We will discuss later in this chapter how this may relate to myocardial preconditioning.

Structurally distinct compounds which were originally described for their ability to relax smooth muscle have been found to open K_{ATP} selectively (see review by Edwards and Weston, 1993). These compounds are characterized by the broad chemical classes of cyanoguanides (pinacidil), benzopyran analogues (cromakalim) and thioformamides (aprikalim) (see *Figure 1*). These compounds open K_{ATP} in smooth muscle and cause relaxation, presumably due to hyperpolarization (Cook, 1988; Ito *et al.*, 1992; Quast and Cook, 1989). Membrane hyperpolarization will reduce calcium entry through voltage-gated channels and thereby relax smooth muscle, although additional mechanisms such as alterations in intracellular calcium release have been proposed. In addition to activating K_{ATP} in smooth muscle, the K_{ATP} openers listed above open K_{ATP} in cardiac myocytes. Since opening of K_{ATP} causes an enhanced outward or repolarizing current, action potentials (AP) are shortened. K_{ATP} openers are not particularly effective at hyperpolarizing cardiac myocytes due to their high resting potential, although they may have some effects on membrane potential when myocytes are depolarized as seen during ischaemia.

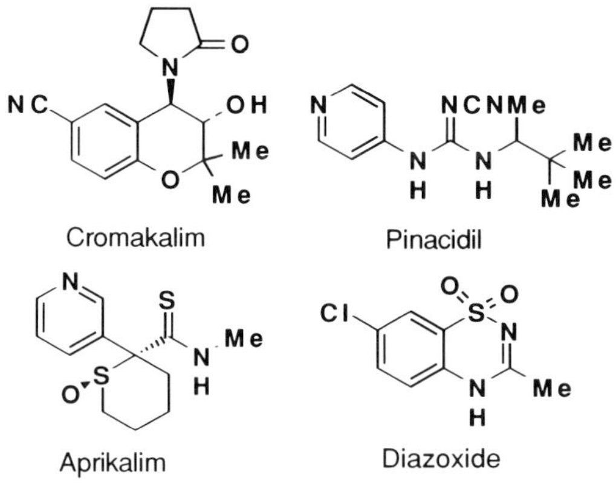

Figure 1. Chemical structures of representatives of various classes of K_{ATP} openers.

AP shortening has long been known to occur during ischaemia. This early repolarization is thought to contribute to the ST-segment shifts (ECG) observed during ischaemia and is caused by an enhanced potassium current (Kantor *et al.*, 1990; Venkatesh *et al.* 1991). Several studies have shown that K_{ATP} may contribute to this enhanced repolarizing current during ischaemia (Kubota *et al.*, 1993; Wilde *et al.*, 1990). ST-segment shifts as well as myocardial potassium efflux are correlated with severity of ischaemia, and it was therefore thought that openers of K_{ATP} might be pro-ischaemic as well as pro-arrhythmic. Conversely, shortening of AP duration may cause a reduced influx of calcium into ischaemic myocytes and therefore K_{ATP} openers were hypothesized by some to be protective. We were fortunate 6–7 years ago to have access to selective K_{ATP} openers in order to determine the role of K_{ATP} in the pathogenesis of myocardial ischaemia.

3.2. *Cardioprotective effects of K_{ATP} openers* in vitro

The first studies using selective K_{ATP} openers in models of myocardial ischaemia were performed in our laboratories (Grover *et al.*, 1989, 1990a). We chose to test them in an isolated rat heart model of global ischaemia and reperfusion in order to avoid the haemodynamic consequences of systemically administering potent vasodilators into intact animals. We first determined the effects of pinacidil and cromakalim in this model of 25 min global ischaemia followed by 30 min of reperfusion (Grover *et al.*,

1989). In this model, there is a poor recovery of contractile function and significant lactate dehydrogenase (LDH) release during reperfusion. Cromakalim and pinacidil significantly enhanced post-ischaemic recovery of function and significantly reduced LDH release. These protective effects were observed within the 1–10 μM range, with no further increase in protection at higher concentrations. The cardioprotective effects of these agents were completely abolished by the K_{ATP} blocker glyburide. Interestingly, glyburide alone did not affect the severity of ischaemia, perhaps because K_{ATP} are not open under the conditions of this model. In *Figure 2*, representative data are shown for the protective effects of cromakalim in isolated rat hearts exposed to global ischaemia and reperfusion. In addition to glyburide, we found that sodium 5-hydroxydecanoate completely abolished the cardioprotective effects of agents such as cromakalim (*Figure 3*).

Further evaluation showed that cromakalim increased the time to the onset of contracture in a concentration-dependent manner during global ischaemia in isolated rat hearts, suggesting a protective effect occurring during ischaemia *per se* (Grover *et al.*, 1991). A potency value was calculated for cardioprotective activity and was denoted as the

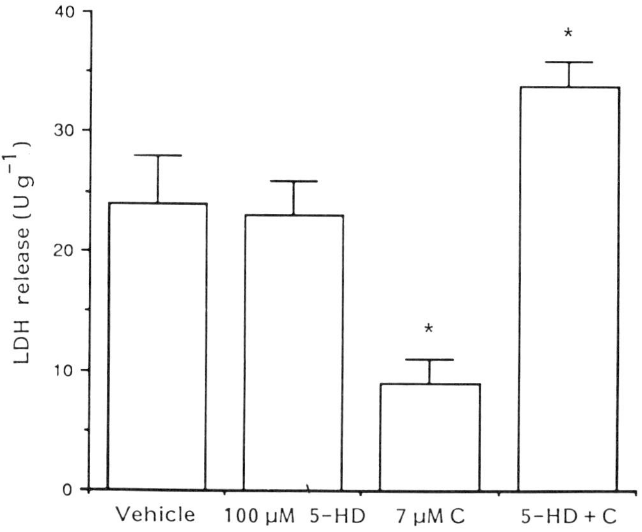

Figure 2. Effect of increasing concentrations of cromakalim on pre-ischaemic and post-ischaemic left ventricular developed pressure (LVDP) in isolated rat hearts made ischaemic for 25 min and reperfused for 30 min. Cromakalim had little effect on pre-ischaemic contractile function but increased the post-ischaemic recovery of function. Reproduced from Grover *et al.* (1990d) with permission from Raven Press.

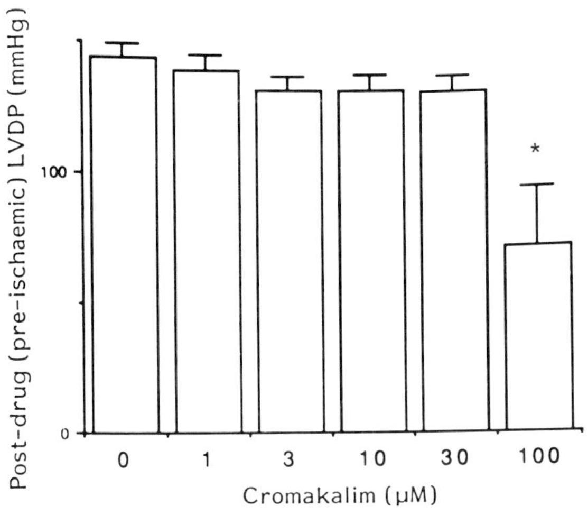

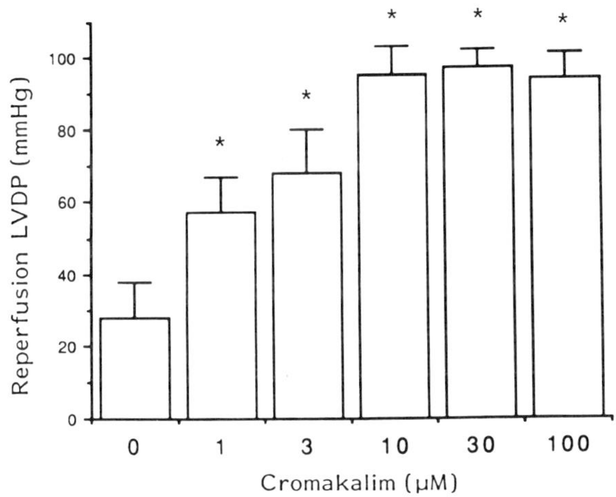

Figure 3. Effect of cromakalim (C) on reperfusion LDH release in isolated rat hearts subjected to 25 min global ischaemia and 30 min reperfusion. Cromakalim significantly reduced LDH release compared to vehicle (*, $p < 0.05$) and this protective effect was completely abolished by the K_{ATP} blocker sodium 5-hydroxydecanoate (5-HD). Data taken from Grover *et al.* (1990d) with permission from Raven Press.

concentration of drug causing a 25% increase in the time to onset of contracture during global ischaemia (EC_{25}). An EC_{25} value of approximately 5 μM has been consistently found for cromakalim. The cardioprotective effects of cromakalim are stereoselective with the (–)-enantiomer (levcromakalim) being the most active (Grover *et al.*, 1991). The cardioprotective effects of cromakalim were lost when given only during reperfusion in the isolated rat heart model (Grover *et al.*, 1990a). Similar profiles of activity have been seen for structurally distinct K_{ATP} openers such as aprikalim and bimakalim with their EC_{25} values being between 1 and 10 μM (Grover *et al.*, 1990b; Sargent *et al.*, 1992). P-1075 (cyanoguanidine) is the one exception, with its EC_{25} being between 30 and 50 nM (Sargent *et al.*, 1993). This is in excellent agreement with its extraordinarily high vasorelaxant potency.

Most other investigators also show cardioprotective effects for K_{ATP} openers in models of myocardial ischaemia *in vitro*. Cole *et al.* (1991) showed that pinacidil improved post-ischaemic recovery of contractile function in a guinea pig perfused ventricle preparation, an effect which was abolished by glyburide. Ohta *et al.* (1991) showed that the K_{ATP} opener KRN2391 enhanced reperfusion function in an isolated rat heart model of ischaemia and reperfusion which was similar to ours. Several studies have shown nicorandil to protect isolated rat hearts, although the protective concentration range is usually in the 100–1000 μM range, which is in agreement with its relatively low K_{ATP} opening potency (relative to other K_{ATP} openers; Grover *et al.*, 1990c; Ohta *et al.*, 1991).

Studies by Pignac *et al.* (1993) showed that aprikalim can exert additional protection to that observed with depolarizing cardioplegic solutions in isolated rabbit hearts. It should be pointed out that the degree of additional protection by aprikalim was small, and this could have been due to the relatively mild ischaemia used. This study was interesting because it was previously unknown whether K_{ATP} openers could protect in the presence of a high potassium solution and also suggests that K_{ATP} openers may work through a mechanism which is distinct from cardioplegia. A study by Galiñanes *et al.* (1992) showed that levcromakalim significantly protected ischaemic/reperfused rabbit hearts, but did not exert additional benefit when given in the presence of depolarizing cardioplegia. These investigators only gave levcromakalim for 1 min with the cardioplegic solution and it was possible that insufficient time was allowed for adequate drug penetration.

3.3. Protective effects of K_{ATP} openers *in vivo*

Cardioprotective effects for K_{ATP} openers have also been observed in whole animal models of myocardial ischaemia, but the results have been more variable than those observed in isolated heart models. Nicorandil was shown many years ago to reduce ischaemic/reperfusion damage in dogs (Gross *et al.*, 1987; Pieper and Gross, 1987). In our hands, nicorandil has direct cardioprotective effects at near millimolar concentrations, which could not be achieved with systemic treatment (Grover *et al.*, 1990c). *In vivo*, nicorandil also does not have an identical profile of cardioprotection compared to other K_{ATP} openers (perhaps due to NO donor activity of nicorandil), but K_{ATP} opening probably does account for some of its protective activity (Grover *et al.*, 1990c). In agreement with this, Mizumura and Gross (1995) showed that the cardioprotective effects of nicorandil are blocked by glyburide in dogs.

The use of more specific K_{ATP} openers showed that K_{ATP} activation can protect ischaemic myocardium *in vivo*. We showed cromakalim and pinacidil to reduce infarct size and induce stunning in dogs (D'Alonzo *et al.*, 1992; Grover *et al.*, 1990a,d) although we had to infuse the drugs directly into the coronary circulation to prevent haemodynamic instability. Gross has successfully infused aprikalim intravenously and found protection in canine models of infarction and stunning (Auchampach *et al.*, 1991, 1992a). Infarct size reduction has also been shown recently for K_{ATP} openers in a porcine model of infarction (Rohmann *et al.*, 1994). While K_{ATP} openers have been found to be efficacious in many *in vivo* models of ischaemia, several studies have been published in which no protective effects were found (Imai *et al.*, 1988; Kitzen *et al.*, 1992; Thornton *et al.*, 1993). Possible reasons for these findings have been suggested to be coronary steal, haemodynamic alterations and the confounding effects of different anaesthetics.

3.4. Mechanism of cardioprotective effects of K_{ATP} openers

An interaction with a cardiac K_{ATP} is probably essential for the cardioprotective effects of K_{ATP} openers. Structurally diverse blockers uniformly abolish the cardioprotective effects of K_{ATP} openers (Cole *et al.* 1991; Grover *et al.*, 1989; Ohta *et al.*, 1991). Glyburide abolishes not only the cardioprotective effects of K_{ATP} openers, but also their coronary vasodilator effects (Grover *et al.*, 1989). Interestingly, glyburide appears to lose some of its blocking activity under ischaemic conditions (Venkatesh *et al.*, 1991). Sodium 5-hydroxydecanoate abolishes the cardioprotective effects of K_{ATP} openers, but this blocking action is activated selectively under ischaemic conditions (McCullough *et al.*, 1991). Meclofenamate also selectively blocks the cardioprotective effects of K_{ATP} openers (independent of cyclo-oxygenase inhibition) with a profile which is similar to sodium 5-hydroxydecanoate (Grover *et al.*, 1994).

We feel that K_{ATP} openers protect ischaemic myocardium directly and this effect is not dependent on vasodilator activity. Isolated hearts are protected by K_{ATP} openers even under constant flow conditions (Grover *et al.*, 1990b). Recently, pyranyl cyanoguanidines have been developed which retain glyburide-reversible cardioprotective effects, but are virtually devoid of vasodilator activity, futher suggesting a direct cardioprotective effect (Atwal *et al.*, 1993). In a recently published study, Armstrong *et al.* (1995) found that pinacidil at 50 μM exerted protective effects on isolated rabbit cardiomyocytes exposed to an ischaemic insult, representing strong evidence of a direct protective effect of K_{ATP} openers on myocytes. Not surprisingly, the cardioprotective effects of pinacidil in this model were abolished by glyburide.

This direct protective effect is also not dependent on reduction of cardiac work as K_{ATP} openers have little cardiodepressant effects at protective concentrations. Some of the protective effects of K_{ATP} openers occur during ischaemia *per se* as time to the onset of contracture during ischaemia is increased. K_{ATP} openers given only during reperfusion are generally not effective, although this does not completely rule out a direct effect on reperfusion injury. Contracture is thought to be due to rigor bond formation secondary to ATP depletion. Consistent with this hypothesis, it was found that K_{ATP} openers conserved ATP during ischaemia, despite a lack of cardiodepressant activity (Grover *et al.*, 1991; McPherson *et al.*, 1993). This suggests a unique profile of cardioprotection because of the separation of energy conservation from effects on cardiac function. It is possible that energy is more efficiently utilized in the presence of K_{ATP} openers and some data suggest that this might be true (Grover *et al.*, 1990d).

An early hypothesis for cardioprotection was that AP shortening by K_{ATP} openers would reduce calcium influx and therefore reduce ischaemic/reperfusion damage. Recent studies have suggested that cardioprotection by K_{ATP} openers is not correlated with AP shortening. Yao and Gross (1994a) showed that a dose of bimakalim could be found which reduced infarct size in dogs, but was without effect on monophasic AP duration. Subsequently, we showed a clear dissociation between cardioprotection and AP shortening for cromakalim in dogs (Grover *et al.*, 1995a). In addition, we have synthesised agents which retain glyburide-reversible cardioprotective effects while being devoid of AP shortening activity using intracellular recording techniques (Grover *et al.*, 1995c). Therefore, increasing evidence suggests that cardioprotection is not correlated with an enhanced sarcolemmal potassium current. Current hypotheses suggest the possibility that an intracellular K_{ATP} is mediating cardioprotection. A K_{ATP} has been shown to be expressed in mitochondrial membranes (Inoue *et al.*, 1991) and this channel is thought to control mitochondrial volume and energetics (Paucek *et al.*, 1992). K_{ATP} openers have been shown to open this mitochondrial K_{ATP} at concentrations which are similar to those resulting in cardioprotection (Paucek *et al.*, 1995).

3.5. Comparison of K_{ATP} openers to preconditioning

K_{ATP} openers exert direct cardioprotective effects which are similar to preconditioning in many ways. Cardiodepression is not necessary for the cardioprotective activity of either K_{ATP} openers or preconditioning. Despite this, ATP levels are conserved during ischaemia with both preconditioning and K_{ATP} openers. The protective effects of preconditioning and K_{ATP} openers are also not dependent on alterations in coronary collateral blood flow.

An important study by Yao and Gross (1994b) showed that bimakalim reduces the threshold for preconditioning in dogs. They found that a combination of a subthreshold dose of bimakalim with a subthreshold preconditioning protocol (3 min preconditioning before 60 min ischaemia) caused significant cardioprotection. This is strong evidence that K_{ATP} openers and preconditioning act through similar mechanisms. If one assumes that some degree of ischaemic insult is necessary to trigger preconditioning, then the addition of a cardioprotective agent might be expected to reduce the efficacy of preconditioning unless similar mechanisms are involved.

There are data suggesting that K_{ATP} openers can stimulate a potassium current which is enhanced with repeated drug exposure. This is similar to the window of 'memory' observed for preconditioning. An interesting example of this is shown by the results of Escande *et al.* (1989) in which pinacidil enhanced steady state currents in guinea pig myocytes more effectively with repeated drug exposure. In between treatments, pinacidil was allowed to wash out.

4. Effect of K_{ATP} blockade on preconditioning

Thus far into the chapter, we have established that K_{ATP} openers mimic preconditioning in many ways, although this alone is not conclusive proof of the importance of K_{ATP} in preconditioning. If K_{ATP} activation is essential for preconditioning, then specific blockers of this channel should inhibit preconditioning. Previous studies from our laboratory showed that K_{ATP} blockers specifically abolish the cardioprotective effects of K_{ATP} openers (Sargent *et al.*, 1991), suggesting that they are not merely non-specific inhibitors of cardioprotection. We are fortunate in having available several structurally distinct blockers. This is particularly important since glyburide may lose some of its efficacy during ischaemia.

Gross's laboratory was the first to show that K_{ATP} blockers can abolish preconditioning. In their first study, they used a model of 5 min of coronary occlusion followed by 60 min of prolonged ischaemia in the same vascular bed (Gross *et al.*, 1992). The profound reduction in infarct size observed in this model of preconditioning was completely abolished by glyburide at a dose which had no effect on infarct size when given alone. The same group performed an additional study in which sodium 5-hydroxydecanoate was used to block K_{ATP} and similar results were found (Auchampach *et al.*, 1992b). Sodium 5-hydroxydecanoate was useful because it can be infused intravenously without effects on plasma glucose. In this same study, these investigators abolished preconditioning with intracoronary glyburide and this was true whether it was given before or after the triggering preconditioning ischaemic episode that preceded prolonged ischaemia (see *Figure 4*).

After these initial studies, several investigators have shown that K_{ATP} blockers inhibit preconditioning. Studies from our laboratory confirmed the results from Gross's group by showing that intracoronary glyburide abolishes preconditioning and this was done in a model which was virtually identical to the one used by Gross's laboratory (Grover *et al.*, 1992). In a recent study, Rohman *et al.* (1994) found that glyburide abolished preconditioning (as measured by infarct size) in a porcine model of preconditioning. In this model, the investigators used 10 min of ischaemia to precondition the hearts followed by a prolonged 60 min ischaemic period. Using a similar porcine model, Schulz *et al.* (1994) found that glyburide completely abolished preconditioning. In addition, they showed that preconditioning slightly enhanced AP shortening during the first minutes of the prolonged ischaemia, although this effect was small (6–7% compared to non-preconditioned animals). Interestingly, these investigators found that while glyburide abolished preconditioning in pigs, it had no effect on adenosine release, suggesting a dissociation between protection and adenosine release. We will discuss adenosine in more detail later in the chapter.

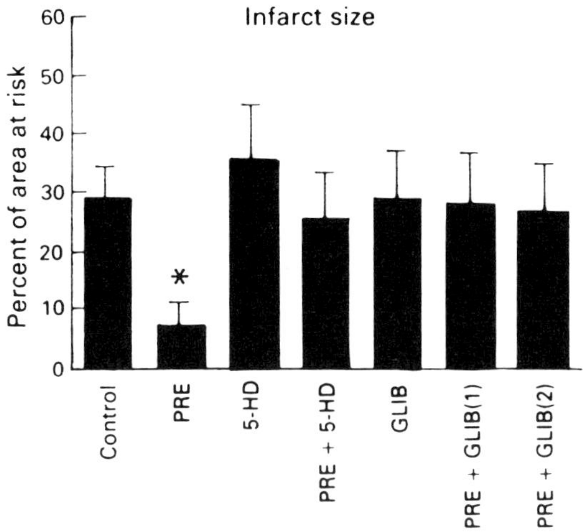

Figure 4. The effect of the K_{ATP} blockers sodium 5-hydroxydecanoate (5-HD) and glyburide (GLIB) on preconditioning in dogs. Five minutes of preconditioning followed by 60 min ischaemia and 5 h reperfusion significantly reduced infarct size. This protective effect was abolished by 5-HD and glyburide whether given during (GLIB1) or after preconditioning (GLIB2). Data taken from Auchampach *et al.* (1992b) with permission from Elsevier Science.

Several studies have shown that preconditioning can be observed in man, further increasing our interest in determining the mechanism of preconditioning. While it is not easy to perform experimental studies in man, glyburide is a useful tool because it is an approved drug which is utilized for treating type II diabetes. Tomai *et al.* (1994) preconditioned patients with a 2 min period of ischaemia induced by inflation of an intracoronary balloon. This was followed by a 5 min period of recovery and a second 2 min coronary occlusion. These investigators found that preconditioning significantly reduced ST-segment shifts associated with subsequent coronary occlusion. A group of patients were also treated with glyburide. Glyburide did not alter ST-segment deviation during the first 2 min coronary occlusion, but abolished the protective effect of preconditioning on the second occlusion. Further evidence for a role of K_{ATP} in preconditioning in man was shown in a study from Yellon's laboratory (Speechly-Dick *et al.*, 1995). This laboratory has developed a model of preconditioning using human atrial trabecula excised from surgical patients. They subjected the tissue to a short period of hypoxia and reoxygenation which was followed by a prolonged hypoxia and subsequent reoxygenation. They observed a poor recovery of contractile function in non-preconditioned tissue while preconditioning significantly enhanced recovery of function. The protective effects of preconditioning in this model were abolished by glyburide. These investigators also showed that cromakalim was protective within the same concentration range as was seen in isolated rat hearts, indicating that K_{ATP} openers can mimic preconditioning in man.

While the studies on the role of K_{ATP} in preconditioning in dogs, pigs and man appear to show straightforward conclusions, some animal models have shown more variable results. Toombs *et al.* (1993a) showed that preconditioning in rabbit hearts could be abolished by K_{ATP} blockers. In a similar *in situ* model of preconditioning, Thornton *et al.* (1993) could not show glyburide to abolish preconditioning. The only recognizable difference between the two models was the anaesthesia employed, with Toombs *et al.* using ketamine–xylazine and Thornton *et al.* using pentobarbital. When the latter laboratory switched to ketamine–xylazine, glyburide was then found to abolish preconditioning (Walsh *et al.*, 1994). No explanation is apparent for the differences observed and it is difficult to reconcile the fact that preconditioning is seen in both models, but that the mechanism is different depending on anaesthesia. Tan *et al.* (1993) determined the effect of preconditioning on the rate of cellular uncoupling in rabbit papillary muscles. The papillary muscles were preconditioned with 10 min of ischaemia followed by a prolonged period of ischaemia. Uncoupling occurred at 15.0 ± 0.7 min in sham muscles and at 22.8 ± 1.5 min in preconditioned tissue. In this particular rabbit model, glyburide was found to abolish the protective effect of preconditioning. These investigators also measured AP duration and found that preconditioning enhanced AP shortening during the prolonged ischaemic episode. This suggests that preconditioning can increase an outward potassium current and this current was found to be blocked by glyburide, further implicating K_{ATP} in the mechanism of preconditioning. Recent data, as described earlier, have suggested that AP shortening may not be correlated with cardioprotection and therefore the significance of AP shortening is presently unclear. A study by Armstrong *et al.* (1995) showed that rabbit cardiomyocytes can be preconditioned and this protective effect could be abolished by glyburide when given during the preconditioning period. Glyburide did not abolish preconditioning when given after the preconditioning period. The authors concluded that K_{ATP}

was necessary for the initiation but not the maintenance of preconditioning. Auchampach *et al.* (1992b) found that glyburide abolished preconditioning even when it was given after preconditioning, although this was in an *in vivo* dog model.

Further confusion has arisen from studies on preconditioning in rats. Rat hearts can be successfully preconditioned, but the mechanism may be different from some other species (Fralix *et al.*, 1993; Grover *et al.*, 1993; Li and Kloner, 1993). Studies from several laboratories, including our own, have shown that glyburide does not abolish preconditioning in isolated rat heart models (Fralix *et al.*, 1993; Grover *et al.*, 1993). An early hypothesis for these findings was that glyburide loses some of its K_{ATP} blocking efficacy under ischaemic conditions and perhaps was not an ideal agent to use in this model (Venkatesh *et al.*, 1991). We repeated these studies with sodium 5-hydroxydecanoate and also found this K_{ATP} blocker to be without effect on preconditioning in isolated rat hearts, despite the fact that this agent is ischaemia-selective (Grover *et al.*, 1995b). The mechanism of preconditioning in rat hearts is still not clear, although results from some laboratories suggest that catecholamines may be mediating preconditioning, possibly via α-adrenoceptor-triggered PKC activation (Banerjee *et al.*, 1993). Lazdunski's laboratory has recently published a paper showing that rat cerebrum can be effectively preconditioned and this preconditioning effect is abolished by glyburide (Herteaux *et al.*, 1995). This suggests that K_{ATP} is either involved with preconditioning in some organs in the rat (tissue differences?) or we are not performing the cardiac studies under the appropriate conditions.

The studies by Lazdunski's group described above suggested that a K_{ATP} involvement in preconditioning can be seen in tissue other than heart. K_{ATP} is richly expressed in skeletal muscles and some of the recent results of Pang *et al.* (1995) indicate an important role of K_{ATP} in this tissue. They preconditioned pig latissimus dorsi muscles with 10 min of global ischaemia followed by 20 min of recovery. A prolonged (4 h) period of ischaemia was then followed by 48 h of reperfusion. Preconditioning significantly reduced infarct size in the latissimus dorsi muscles, an effect which was simulated by levcromakalim and adenosine. Preconditioning was completely abolished by glyburide and sodium 5-hydroxydecanoate.

5. Where does K_{ATP} fit into the sequence of events during preconditioning?

5.1. Interaction of adenosine A_1 receptors and K_{ATP}

Early studies showed the involvement of adenosine A_1 receptors in the mechanism of preconditioning. Studies by Kirsch *et al.* (1990) suggested that adenosine A_1 receptor stimulation can increase K_{ATP} open probability in rat cardiomyocytes and this occurs through a G_i protein-coupled mechanism. It therefore seemed logical that preconditioning-induced adenosine release might open K_{ATP} and afford cardioprotection. With this in mind, we determined the effect of the adenosine A_1 receptor agonist R-PIA [N^6-1-(phenyl-2*R*-isopropyl)adenosine] on infarct size in dogs and the effect of glyburide on its activity (Grover *et al.*, 1992). We found R-PIA to reduce infarct size, with this effect being completely abolished by glyburide (see *Figure 5*). Van Winkle *et al.* (1994) showed that the cardioprotective effects of R-PIA in a porcine model of infarction were abolished by sodium 5-hydroxydecanoate. Adenosine has also been

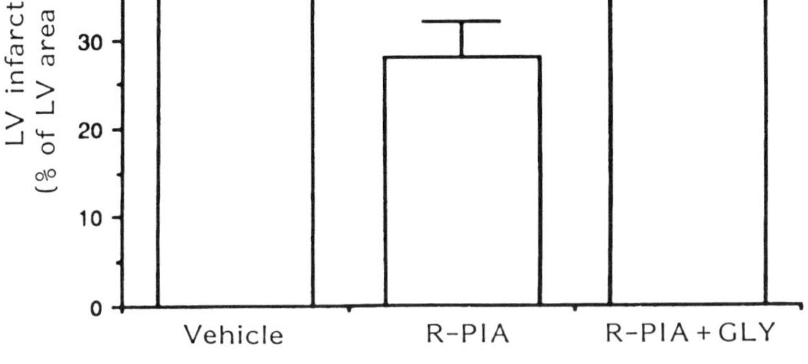

Figure 5. The effect of R-PIA (adenosine A_1 agonist) on left ventricular (LV) infarct size in dogs subjected to 90 min coronary occlusion followed by 5 h reperfusion. R-PIA significantly reduced infarct size and this effect was completely abolished by glyburide (GLY).

shown to protect ischaemic dog and rabbit hearts and this protective effect is abolished by glyburide (Auchampach and Gross, 1993; Toombs *et al.*, 1993b).

While these results suggest that adenosine release activates K_{ATP}, other studies have suggested a reverse sequence. A study by Tsuchida *et al.* (1993) showed that pinacidil could protect ischaemic/reperfused rabbit myocardium, but interestingly, this effect was abolished by the adenosine receptor antagonist SPT (8-*p*-sulphophenyl theophylline). A study by Armstrong *et al.* (1995) showed that pinacidil protected ischaemic isolated rabbit cardiomyocytes and this effect was abolished by SPT, DPCPX and adenosine deaminase. A recent study by Kitakaze *et al.* (1994) suggested that K_{ATP} openers are protecting ischaemic myocardium by increasing adenosine production secondary to activation of 5′-nucleotidase. It is difficult to reconcile the differences between the various groups as to whether K_{ATP} opening leads to adenosine production or the reverse. Studies from Gross' and our laboratories have shown that the infarct sparing effect of bimakalim are not abolished by DPCPX in dogs and rats (Gross *et al.*, 1995). In rats we also could not show glyburide to abolish the protective effects of R-PIA nor could we show DPCPX to abolish the protective effects of cromakalim (Grover *et al.*, 1996b). This suggests there is no pharmacological link between K_{ATP} and adenosine A_1 receptors in rat hearts, which is ironic since Kirsch's original study establishing such a link was carried out in neonatal rat myocytes. This lack of a pharmacological link between K_{ATP} and adenosine receptors may explain the apparent differences noted for preconditioning in rat hearts.

5.2. Mechanism of K_{ATP}-induced cardioprotection in preconditioning: what is the end-effector?

Several potentially important pathways have been suggested to be involved in preconditioning. It is unclear how these various pathways fit together and it is even less clear what is the ultimate end-effector. Recently, several investigators have suggested that PKC activation may be an important step in mediating preconditioning, although it is not known which protein is phosphorylated by this enzyme (Banerjee *et al.*, 1993). Recent studies from Yellon's laboratory have suggested that PKC may interact with K_{ATP} to afford cardioprotection. As described earlier, this group showed that it is possible to precondition human atrial trabecula using post-ischaemic functional recovery as an index of damage (Speechly-Dick *et al.,* 1995). These investigators found cromakalim to protect this tissue, whilst glyburide abolished the protective effects of both preconditioning and cromakalim. Interestingly, they found that glyburide also abolished the protective effects of PKC activators. From these data, they suggested that PKC may phosphorylate a protein which is directly or indirectly linked to K_{ATP}. While these data indicate an important role for K_{ATP} in preconditioning and that these channels are 'downstream' of PKC activation, it still does not prove that K_{ATP} is the end-effector. Unfortunately, we do not understand the cascade of biochemical events occurring upon K_{ATP} opening. Until we have an improved understanding of how K_{ATP} openers protect ischaemic myocardium or how blockers abolish preconditioning, it will be difficult to determine how K_{ATP} can exert protection during preconditioning. At the present time, the best means at our disposal is to use pharmacological K_{ATP} openers for determining mechanisms. It would also be extremely useful to know the binding site of the sulphonylureas and how this site interacts with the cardioprotective site. We will review what is known and suggest future directions.

As previously described, the action potential shortening of K_{ATP} openers does not correlate with cardioprotection (Grover *et al.,* 1995a; Yao and Gross, 1994a). In our study, the cardioprotective effects of cromakalim were observed in the presence of a dose of the delayed rectifier blocker dofetilide which abolished the AP shortening action of cromakalim. Unpublished data from our laboratories have also shown that dofetilide (abolished AP shortening) does not abolish preconditioning, suggesting a lack of correlation between AP shortening and the cardioprotection caused by preconditioning. These data are reassuring since the class III anti-arrhythmic agents will not be expected to abolish preconditioning. We therefore hypothesize that sarcolemmal K_{ATP} are not involved in the mechanism of preconditioning or in the cardioprotective mechanism of K_{ATP} openers. Further confirmation of this hypothesis comes from our findings that some pyranyl cyanoguanidine analogues (structurally related to both pinacidil and cromakalim) retain glyburide-reversible cardioprotective effects despite being devoid of vasodilator and AP shortening activity, both of which are associated with sarcolemmal K_{ATP} opening (Atwal *et al.,* 1993; Grover *et al.,* 1996a). Based on our knowledge of the pharmacology of K_{ATP} openers, we have hypothesized an intracellular site of action. As discussed earlier, a K_{ATP} has been found on mitochondrial membranes and it is thought to regulate mitochondrial volume and energetics (Inoue *et al.*, 1991; Paucek *et al.*, 1992). Recently, Garlid's group has shown that selected K_{ATP} openers can activate this mitochondrial K_{ATP}, which is consistent with our hypothesis, but does not prove it (Paucek *et al.*, 1995). A critical study would be to determine whether the 'selective' pyranyl cyanoguanidine analogues such as BMS-180448 (Grover *et al.*, 1996a) can open mitochondrial K_{ATP} despite a lack of effect on sarcolemmal K_{ATP}. This would be further proof that mitochondrial K_{ATP} are involved. If BMS-180448 can open this mitochondrial channel via a stereoselective and glyburide-reversible mechanism, we will be well on our way to determining the cardioprotective site of action of K_{ATP} openers.

At the present time, the relevant cardioprotective binding site for K_{ATP} openers is not known. It would also be very useful to determine the protein binding site in order to better understand their mechanism of action, and presumably this would also shed light on the mechanism of preconditioning. It is felt that sulphonylureas do not displace K_{ATP} openers from a common binding site, but apparently these receptors are allosterically linked (Aguilar-Bryon *et al.*, 1995; Phillipsone and Steiner, 1995). Since glyburide is an important tool for determining the role of K_{ATP} in preconditioning, it would also be useful to learn more about its site of action. As previously described, a sulphonylurea binding site (SUR) has been found to be a member of the ATP-binding cassette protein family and this protein has no apparent ion permeant properties (Al-Aqati, 1995). It is thought that SUR secretes ATP which then interacts with purinergic receptors and increases K_{ATP} opening probability. Most of this work has been done in insulin-secreting cells and it remains to be proven if SUR is relevant to the cardioprotective effects of preconditioning in the heart.

6. Conclusions

There is a growing interest in the role of K_{ATP} in the pathogenesis of myocardial ischaemia. The majority of data collected thus far suggest that K_{ATP} opening is protective in models of myocardial ischaemia. The intriguing work linking K_{ATP} activation with the mechanism of preconditioning suggests that K_{ATP} may serve as an endogenous protective mechanism. While most studies conducted suggest a role for K_{ATP} in preconditioning, we are a long way from understanding how such a mechanism can protect hearts. Only after we learn more about the molecular mechanism of protection of K_{ATP} activation will we be positioned to understand precisely where and how K_{ATP} fits into the cascade of events of preconditioning. There is much effort to determine the end-effector of preconditioning, and at the present time we do not know whether K_{ATP} represents this end-effector. There is great hope that the work on the mechanism of preconditioning could lead to novel therapeutics. Design of such therapeutics may not be centred around the end-effector because it may not lend itself to selective modulation. Whether K_{ATP} is the end-effector or not, we believe that pharmacological activation of K_{ATP} represents a viable target. Determination of the molecular mechanism of action of currently existing agents could lead to the design of more selective and potent compounds, which may better mimic preconditioning.

References

Aguilar-Bryan L, Nichols CG, Wechsler SW *et al.* (1995) Cloning of the beta-cell high-affinity sulfonylurea receptor: a regulator of insulin secretion. *Science,* **268,** 423–425.

Al-Aqati Q. (1995) Regulation of ion channels by ABC transporters that secrete ATP. *Science,* **269,** 805–806.

Armstrong SC, Liu G, Downey J and Ganote C. (1995) K_{ATP} channels and preconditioning of rabbit cardiomyocytes. *J. Mol. Cell Cardiol.* **27,** A23.

Atwal KS, Grover GJ, Ahmed S *et al.* (1993) Cardioselective anti-ischemic ATP-sensitive potassium channel openers. *J. Med. Chem.* **36,** 3971–3974.

Auchampach JA and Gross GJ. (1993) Adenosine A_1 receptors, K_{ATP} channels and ischemic preconditioning in dogs. *Am. J. Physiol.* **264,** H1327–1336.

Auchampach JA, Maruyama M, Cavero I and Gross GJ. (1991) The new K^+ channel opener aprikalim (RP 52891) reduces experimental infarct size in dogs in the absence of hemodynamic changes. *J. Pharmacol. Exp. Ther.* **259,** 961–967.

Auchampach JA, Maruyama M, Cavero I and Gross G. (1992a) Pharmacologic evidence for a role of ATP-dependent potassium channels in myocardial stunning. *J. Pharmacol. Exp. Ther.* **86,** 311–319.

Auchampach JA, Grover GJ and Gross GJ. (1992b) Blockade of ischemic preconditioning in dogs by the novel ATP dependent potassium channel antagonist sodium 5-hydroxydecanoate. *Cardiovasc. Res.* **26,** 1054–1062.

Banerjee A, Locke-Winter C, Rogers KB, Mitchel MB, Brew EC, Cairns CB, Bensard DD and Harken AH. (1993) Preconditioning against myocardial dysfunction after ischemia and reperfusion by an $alpha_1$-adrenergic mechanism. *Circ. Res.* **73,** 656–670.

Chen W, Gabel S, Steenbergen C and Murphy E. (1995) A redox-based mechanism for cardioprotection induced by ischemic preconditioning in perfused rat heart. *Circ. Res.* **77,** 424–429.

Cole WC, McPherson CD and Sontag D. (1991) ATP-regulated channels protect the myocardium against ischemia/reperfusion damage. *Circ. Res.* **69,** 571–581.

Cook NS. (1988) The pharmacology of potassium channels and their therapeutic potential. *Trends Pharmacol.* **9,** 21–28.

D'Alonzo AJ, Darbenzio RB, Parham CS and Grover GJ. (1992) Effects of intracoronary cromakalim on postischemic contractile function and action potential duration. *Cardiovasc. Res.* **26,** 1046–1053.

Edwards G and Weston AH. (1993) The pharmacology of ATP-sensitive potassium channels. *Ann. Rev. Pharmacol. Toxicol.* **33,** 597–637.

Escande D, Thuringer D, Le Guern S, Courteix J, Laville M and Cavero I. (1989) Potassium channel openers act through an activation of ATP-sensitive K^+ channels in guinea-pig cardiac myocytes. *Plugers Arch.* **414,** 669–675.

Fralix TA, Steenbergen C, London RE and Murphy E. (1993) Glibenclamide does not abolish the protective effect of preconditioning on stunning in the isolated perfused rat heart. *Cardiovasc. Res.* **27,** 630–637.

Galiñanes M, Shattock MJ and Hearse DJ. (1992) Effects of potassium channel modulation during global ischaemia in isolated rat heart with and without cardioplegia. *Cardiovasc. Res.* **26,** 1063–1068.

Gross GJ and Auchampach JA. (1992) Blockade of ATP-sensitive potassium channels prevents myocardial preconditioning in dogs. *Circ. Res.* **70,** 223–233.

Gross GJ, Peiper GM and Warltier DC. (1987) Comparative effects of nicorandil, nitroglycerin, nicotinic acid, and SG-86 on the metabolic status and functional recovery of the ischemic-reperfused myocardium. *J. Cardiovasc. Pharmacol.* **10** (Suppl. 8), S76–84.

Gross GJ, Sleph PG and Grover PG. (1995) Cardioprotective effects of K_{ATP} openers occur independently of adenosine A_1 receptor activation. *J. Mol. Cell Cardiol.* **27,** A24.

Grover GJ, McCullough JR, Henry DE, Conder ML and Sleph PG. (1989) Anti-ischemic effects of the potassium channel activators pinacidil and cromakalim and the reversal of these effects with the potassium channel blocker glyburide. *J. Pharmacol. Exp. Ther.* **251,** 98–104.

Grover GJ, Dzwonczyk S, Parham CS and Sleph PG. (1990a) The protective effects of cromakalim and pinacidil on reperfusion function and infarct size in isolated perfused rat hearts and anesthetized dogs. *Cardiovasc. Drugs Ther.* **4,** 465–74.

Grover GJ, Dzwonczyk S and Sleph PG. (1990b) Reduction of ischemic damage in isolated rat hearts by the potassium channel opener RP 52891. *Eur. J. Pharmacol.* **191,** 11–19.

Grover GJ, Sleph PG and Parham CS. (1990c) Nicorandil improves postischemic contractile function independently of direct myocardial effects. *J. Cardiovasc. Pharmacol.* **15,** 698–705.

Grover GJ, Sleph PG and Dzwonczyk S. (1990d) Pharmacologic profile of cromakalim in the treatment of myocardial ischemia in isolated rat hearts and anesthetized dogs. *J. Cardiovasc. Pharmacol.* **16,** 853–864.

Grover GJ, Newburger J, Sleph PG, Dzwonczyk S, Taylor SC, Ahmed SZ and Atwal KS. (1991) Cardioprotective effects of the potassium channel opener cromakalim, stereoselectivity and effects on myocardial adenine nucleotides. *J. Pharmacol. Exp. Ther.* **257,** 156–162.

Grover GJ, Sleph PG and Dzwonczyk S. (1992) Role of myocardial ATP-sensitive potassium channels in mediating preconditioning in the dog heart and their possible interaction with adenosine A_1-receptors. *Circulation,* **86,** 1310–1316.

Grover GJ, Dzwonczyk S, Sleph PG and Sargent CA. (1993) The ATP-sensitive potassium channel blocker glibenclamide (glyburide) does not abolish preconditioning in isolated ischemic rat hearts. *J. Pharmacol. Exp. Ther.* **265,** 559–564.

Grover GJ, D'Alonzo J, Sleph PG, Dzwonczyk S, Hess TA and Darbenzio RB. (1994) The cardioprotective and electrophysiologic effects of cromakalim are attenuated by meclofenamate through a cyclooxygenase-independent mechanism. *J. Pharmacol. Exp. Ther.* **269,** 536–540.

Grover GJ, D'Alonzo AJ and Parham CS. (1995a) Cardioprotection with the K_{ATP} opener cromakalim is not correlated with ischemic myocardial action potential duration. *J. Cardiovasc. Pharmacol.* **26,** 145–152.

Grover, GJ, Murray HN, Baird A and Dzwonczyk S. (1995b) The K_{ATP} blocker sodium 5-hydroxydecanoate does not abolish preconditioning in isolated rat hearts. *Eur. J. Pharmacol.* **277,** 271–274.

Grover GJ, D'Alonzo AJ, Hess TA and Sleph PG. (1995c) Glyburide-reversible cardioprotective effect of BMS-180448 is independent of action potential shortening in guinea pig hearts. *Cardiovasc. Res.,* **30:** 731–738.

Grover GJ, Baird AJ and Sleph PG. (1996) Lack of a pharmacologic interaction between ATP-sensitive potassium channels and adenosine A_1 receptors in ischemic rat hearts. *Cardiovasc. Res.,* in press.

Heurteaux C, Lauritzen I, Widmann C and Lazdunski M. (1995) Essential role of adenosin, adenosine A_1 receptors, and ATP-sensitive K^+ channels in cerebral ischemic preconditioning. *Proc. Natl Acad. Sci. USA* **92,** 4666–4670.

Imai N, Liang C, Stone CK, Sakamoto S and Hood WB. (1988) Comparative effects of nitroprusside and pinacidil on myocardial blood flow and infarct size in awake dogs

with acute myocardial infarction. *Circulation,* **77,** 705–711.
Inoue I, Nagase H, Kishi K and Higuti T. (1991) ATP-sensitive K^+ channel in the mitochondrial inner membrane. *Nature,* **352,** 244–247.
Ito K, Kanno T, Suzuki K, Masuzawa-Ito K, Takewaki T, Ohashi H, Asano M and Suzuki H. (1992) Effects of cromakalim on the contraction and the membrane potential of the circular smooth muscle of guinea-pig stomach. *Br. J. Pharmacol.* **105,** 335–340.
Jennings RB, Murry CE and Reimer KA. (1991) Preconditioning myocardium with ischemia. *Cardiovasc. Drug Ther.* **5,** 933–938.
Kantor PF, Coetzee WA, Carmeliet EE, Dennis SC and Opie LH. (1990) Reduction of ischemic K^+ loss and arrhythmias in rat hearts. Effect of glybenclamide, a sulfonylurea. *Circ. Res.* **66,** 478–485.
Kirsch CE, Codina J, Birnbaumer L and Brown AM. (1990) Coupling of ATP-sensitive K^+ channels to A_1 receptors by G proteins in rat ventricular myocytes. *Am. J. Physiol.* **258,** H820–H826.
Kitakaze M, Minamino T, Node K, Komamura K, Shinozake Y, Chujo M, Mori H, Hori M and Kamada T. (1994) Opening of K^+ channels mimics the infarct size limiting effect of ischaemic preconditioning, role of activation of ectosolic 5'-nucleotidase. *Eur. Heart J.* **15** (Suppl.), 482.
Kitzen JM, McCallum JD, Harvey C, Morin ME, Oshiro GT and Colatsky TJ. (1992) Potassium channel activators cromakalim and celikalim (WAY-120,491) fail to decrease myocardial infarct size in the anesthetized canine. *Pharmacology,* **45,** 71–82.
Kubota I, Yamaki M, Shibata T, Ikeno E, Hosoya Y and Tomoike H. (1993) Role of ATP-sensitive K^+ channel on ECG ST segment elevation during a bout of myocardia ischemia. A study on epicardial mapping in dogs. *Circulation,* **88,** 1845–1851.
Li Y and Kloner RA. (1993) The cardioprotective effects of ischemic 'preconditioning' are not mediated by adenosine receptors in rat hearts. *Circulation,* **87,** 1642–1648.
Light PE, Allen BG, Walsh MP and French RJ. (1995) Regulation of adenosine triphosphate-sensitive potassium channels from rabbit ventricular myocytes by protein kinase C and type 2A protein phosphatase. *Biochemistry,* **34,** 7252–7257.
McCullough JR, Normandin DE, Conder ML, Sleph PG, Dzwonczyk S and Grover GJ. (1991) Specific block of the anti-ischemic actions of cromakalim by sodium 5-hydroxydecanoate. *Circ. Res.* **69,** 949–958.
McPherson CD, Pierce GN and Cole WC. (1993) Ischemic cardioprotection by ATP-sensitive potassium channels involves high-energy phosphate preservation. *Am. J. Physiol.* **265,** H1809–H1818.
Mizumura T and Gross GJ. (1995) The cardioprotective effect of nicorandil, a K_{ATP} channel opener nitrate, is blocked by glyburide in dogs. *J. Mol. Cell Cardiol.* **27,** A24.
Murry, CE, Richard VJ, Reimer KA and Jennings RB. (1990) Ischemic preconditioning slows energy metabolism and delays ultrastructural damage during a sustained ischaemic episode. *Circ. Res.* **66,** 913–931.
Noma A. (1983) ATP-regulated K^+ channels in cardiac muscle. *Nature,* **305,** 147–148.
Ohta H, Jinno Y, Harada K, Ogawa N, Fukushima and Nishikore J. (1991) Cardioprotective effects of KRN2391 and nicorandil on ischemic dysfunction in perfused rat heart. *Eur. J. Pharmacol.* **204,** 171–177.
Pang CY, Zhong AG, Yang R and Forrest CR. (1995) Involvement of $adenosine_1$ (A_1) receptors and ATP-sensitive K^+ (K_{ATP}) channels in ischemic preconditioning (IPC) of pig skeletal muscles. *FASEB J.* **9,** A422.
Paucek P, Mironova G, Mahdi F, Beavis AD, Woldegiorgis G and Garlid KD. (1992) Reconstitution and partial purification of the glibenclamide-sensitive, ATP-dependent K^+ channel from rat liver and beef heart mitochondria. *J. Biol. Chem.* **36,** 26062–26069.
Paucek P, Yarov-Yarovoy V, Sun X and Garlid KD. (1995) Physiological and pharmacological activators of the mitochondrial K_{ATP} channel. *Biophys. J.* **68,** A145.
Philipsone LH and Steiner DF. (1995) *Pas de deux* or more, the sulfonylurea receptor and K^+ channels. *Science,* **268,** 372–373.
Pieper GM and Gross GJ. (1987) Salutary action of nicorandil, a new antianginal drug, on myocardial metabolism during ischemia and on postischemic function in a canine preparation of brief, repetitive coronary artery occlusion, comparison with isosorbide dinitrate. *Circulation,* **76,** 916–928.

Pignac J, Bourgouin J and Dumont L. (1993) Cold cardioplegia and the K^+ channel modulator aprkalim, improved cardioprotection in isolated ischemic rabbit hearts. *Can. J. Physiol. Pharmacol.* **72,** 126–132.

Quast U and Cook NS. (1989) Moving together, K^+ channel openers and ATP-sensitive K^+ channels. *Trends Pharmacol. Sci.* **10,** 431–435.

Ribalet B, Ciani S and Eddlestone GT. (1989) ATP mediates both activation and inhibition of K_{ATP} channel activity via cAMP-dependent protein kinase in insulin-secreting cell lines. *J. Gen. Physiol.* **94,** 693–717.

Rohmann S, Weygandt H, Schelling P, Kie Soei L, Verdouw PD and Lues I. (1994) Involvement of ATP-sensitive potassium channels in preconditioning protection. *Basic Res. Cardiol.* **89,** 563–576.

Sargent CA, Smith MA, Dzwonczyk S, Sleph PG and Grover GJ. (1991) Effect of potassium channel blockade on the anti-ischemic actions of mechanistically diverse agents. *J. Pharmacol. Exp. Ther.* **259,** 97–103.

Sargent CA, Dzwonczyk S and Grover GJ. (1992) Effect of the potassium channel opener EMD 56431 on globally ischemic rat hearts. *Pharmacology,* **45,** 260–268.

Sargent CA, Dzwonczyk S,Sleph PG, Normandin DE, Antonaccio MJ and Grover GJ. (1993) Cardioprotective effects of the cyanoguanidine potassium channel opener P-1075. *J. Cardiovasc. Pharmacol.* **22,** 564–570.

Schulz R, Rose J and Heusch G. (1994) Involvement of activation of ATP-dependent potassium channels in ischemic preconditioning in swine. *Am. J. Physiol.* **267,** H1341–H1352.

Speechly-Dick ME, Grover GJ and Yellon DM. (1995) Does ischemic preconditioning in the human involve protein kinase C and the ATP-dependent potassium channel? Studies of contractile function in an *in vitro* model. *Circ. Res.* **77,** 1030–1035.

Spruce AE, Standen NB and Stanfield PR. (1985) Voltage-dependent ATP-sensitive potassium channels of skeletal muscle membrane. *Nature,* **316,** 736–738.

Tan HL, Mazon P, Verberne HJ, Sleeswijk ME, Coronel R, Opthof T and Janse MJ. (1993) Ischemic preconditioning delays ischemia induced cellular uncoupling in rabbit myocardium by activation of ATP-sensitive potassium channels. *Cardiovasc. Res.* **27,** 644–651.

Thornton JD, Thornton CS, Sterlin DL and Downey JM. (1993) Blockade of ATP-sensitive potassium channels increases infarct size but does not prevent preconditioning in rabbit hearts. *Circ. Res.* **72,** 44–49.

Tomai F, Crea F, Gaspardone A, Versaci G, De Paulis R, Penta de Peppo A, Chiariello L and Gioffre PA. (1994) Ischemic preconditioning during coronary angioplasy is prevented by glibenclamide, a selective ATP-sensitive K^+ channel blocker. *Circulation,* **90,** 700–705.

Toombs CF, Moore TL and Shebuski RJ. (1993a) Limitation of infarct size in the rabbit by ischemic preconditioning is reversible with glibenclamide. *Cardiovasc. Res.* **27,** 617–622.

Toombs CF, McGee DS, Johnston WE and Vinten-Johansen J. (1993b) Protection from ischaemic-reperfusion injury with adenosine treatment is reversed by inhibition of ATP-sensitive potassium channels. *Cardiovasc. Res.* **27,** 623–629.

Treherne JM and Ashford MLJ. (1991) The regional distribution of sulfphonylurea binding sites in rat brain. *Neuroscience,* **40,** 523–531.

Tsuchida A, Walsh RS and Downey J. (1993) Protection by the ATP-sensitive K^+ opener pinacidil can be blocked with an adenosine receptor antagonist. *Circulation,* **88,** I632.

Van Winkle DM, Chien GL, Wolff RA, Soifer BE, Kuzume K and Davis F. (1994) Cardioprotection provided by adenosine receptor activation is abolished by blockade of the K_{ATP} channel. *Am. J. Physiol.* **266,** H829–H839.

Venkatesh N, Lamp ST and Weiss JN. (1991) Sulfonylureas, ATP-sensitive K^+ loss during hypoxia, ischemia, and metabolic inhibition in mammalian ventricle. *Circ. Res.* **69:** 623–629.

Walsh RS, Tsuchida A, Daly JJF, Thornton JD, Cohen MV and Downey JM. (1994) Ketamine-xylazine anesthesia permits a K_{ATP} channel antagonist to attenuate preconditioning in rabbit myocardium. *Cardiovasc. Res.* **28,** 1337–1341.

de Weille J, Schmid-Antomarchi H, Fosset M and Lazdunski M. (1988) ATP-sensitive K^+ channels that are blocked by hypoglycemia-inducing sulfonylureas in insulin-secreting cells are activated by galanin, a hyperglycemia-inducing hormone. *Proc. Natl Acad. Sci. USA* **85,** 1312–1316.

de Weille J, Schmid-Antomarchi H, Fosset and Lazdunski M. (1989) Regulation of ATP-sensitive K^+ channels in insulinoma cells, activation by somatostatin and protein kinase C and the role of cAMP. *Proc. Natl Acad. Sci. USA* **86,** 2971–2975.

Wilde AAM, Escande D, Schumacher CA, Thuringer D, Mestre M, Fiolet JWT and Janse MJ. (1990) Potassium accumulation in the globally ischemic mammalian heart. A role for the ATP-sensitive potassium channel. *Circ. Res.* **67,** 835–843.

Yao Z and Gross GJ. (1994a) Effects of the K_{ATP} opener bimakalim on coronary blood flow, monophasic action potential duration, and infarct size in dogs. *Circulation,* **89,** 1769–1775.

Yao Z and Gross GJ. (1994b) Activation of ATP-sensitive potassium channels lowers threshold for ischemic preconditioning in dogs. *Am. J. Physiol.* **267,** H1888–H1894.

4

Ischaemic preconditioning and arrhythmias induced by ischaemia and reperfusion

Mark Connaughton[1], Clive S. Lawson[2] and David J. Hearse[3]
[1]Department of Cardiology, Royal Brompton Hospital, Sydney Street, London SW3 6NP, UK
[2]Department of Cardiology, Kent and Sussex Hospital, Mount Ephraim, Tunbridge Wells, Kent, UK
[3]Cardiovascular Research, The Rayne Institute, St Thomas' Hospital, London SE1 7EH, UK

1. The history of preconditioning against arrhythmias

Although the original description of 'preconditioning with ischaemia' was first coined using a model of infarct size limitation in dogs (Murry *et al.*, 1986), only a short interval passed before arrhythmias were used as an alternative experimental endpoint (Shiki and Hearse, 1987). The original report from the Duke University group generated considerable excitement, since it provided the first description of an intervention which could reliably and reproducibly limit infarct size under controlled experimental conditions. In the decade following Murry and colleagues' original description the catchy nomenclature associated with the phenomenon has, as with other recent concepts in cardiac

Ischaemia: preconditioning and adaptation, edited by M.S. Marber and D.M. Yellon

pathophysiology, stimulated a great deal of interest (Connaughton and Hearse, 1996; Przyklenk and Kloner, 1995).

Paradoxically, the recent surge of interest in preconditioning has diverted attention from older descriptions of myocardial responses to serial insults, some of which predate Murry's classic paper by more than a third of a century. Thus, in 1950, Harris (1950) described a two stage coronary ligation protocol (a partial occlusion prior to total occlusion) which caused a significant reduction in the number of ventricular premature beats (VPB) induced early in ischaemia. More than 25 years later, Gulker and colleagues (1977) showed in a canine model, that serial episodes of ischaemia–reperfusion were associated with an increase in the threshold for precipitation of ventricular fibrillation (VF) by programmed electrical stimulation. Six years following this, Barber (1983) reported that repeated 5 min coronary occlusions caused reductions both in the number of VPBs induced by ischaemia, and in the extent of ST segment elevation on the electrocardiogram. At the time, these findings received only limited recognition, possibly because there were many prior reports of effective anti-arrhythmic interventions, and possibly also because the degree of protection was not particularly marked.

Many subsequent studies using protocols now described as 'preconditioning' have shown that profound protection can be induced against both ischaemia- and reperfusion-induced arrhythmias (e.g. Lawson *et al.*, 1993a,b; Li *et al.*, 1992; Shiki and Hearse, 1987; Vegh *et al.*, 1992a). However, protection against arrhythmias in such protocols is by no means universal (Murry *et al.*, 1986) and in some cases the prior ischaemia–reperfusion cycles even appear to be pro-arrhythmic (Reimer *et al.*, 1990). Whether the molecular substrates of any anti-arrhythmic protection induced by preconditioning are the same as those involved in delaying the onset of myocardial necrosis is a question which is not yet clearly resolved.

2. Experimental preconditioning against arrhythmias

2.1. Preconditioning against reperfusion-induced arrhythmias

The first demonstration that preconditioning could actually prevent *malignant* arrhythmias was reported by Shiki and Hearse (1987). They examined the effect of preconditioning on the severity of reperfusion-induced arrhythmias in rat hearts *in vivo* using paired coronary occlusions separated by a variable time period from 10 min to 3 days. VF induced by reperfusion and VPBs following the second ischaemic challenge could be abolished and a profound reduction was noted in the incidence of reperfusion-induced ventricular tachycardia (VT) with a coincident increase in the 'time-in-sinus-rhythm' following reperfusion (*Figure 1*). This study defined the temporal characteristics of the envelope of protection, showing that the preconditioning coronary occlusion had to last at least 3 min, took only 15 min to provide effective protection and that the subsequent challenge had to take place within 60 min of the preconditioning stimulus for the protection to be retained. These temporal parameters showed a striking similarity to those described for preconditioning against myocardial necrosis (Li *et al.*, 1992; Murry *et al.*, 1986), suggesting the possibility that the two might be manifestations of a common protective phenomenon. In molecular terms this relatively rapid onset of action suggested activation of a pre-existing signalling system, rather than dependence on a slower process such as protein synthesis which would have required some hours to take effect. A major insight provided by this study was the recognition of the rapidity of onset of preconditioning-induced protection. Studies using myocardial infarct size as the primary endpoint need several hours of reperfusion before the size of the infarct becomes maximal. In a protocol of this kind it is not therefore possible to determine how quickly protection is manifest. Using arrhythmias as an endpoint allowed the demonstration that protection was effective only 15 min after the first coronary occlusion.

2.2. Increase in ischaemic tolerance or true reduction in arrhythmia severity?

The incidence of reperfusion-induced arrhythmias is critically dependent on the duration of the preceding ischaemia, the relationship between the two being 'bell-shaped' (Manning and Hearse, 1984). Only limited information can therefore be gained from protocols in which only one ischaemic duration is studied. Brief episodes of ischaemia (e.g. <3 min) cause little ischaemic damage and the incidence of reperfusion-induced arrhythmias is low. Similarly, if the ischaemic duration is very prolonged (e.g. >1 h) irreversible injury is inevitable and reperfusion of non-viable myocardium causes few arrhythmias. Between these extremes (e.g. with ischaemic durations of 5–20 min), the incidence of reperfusion-induced malignant arrhythmias is maximal and is highly predictable.

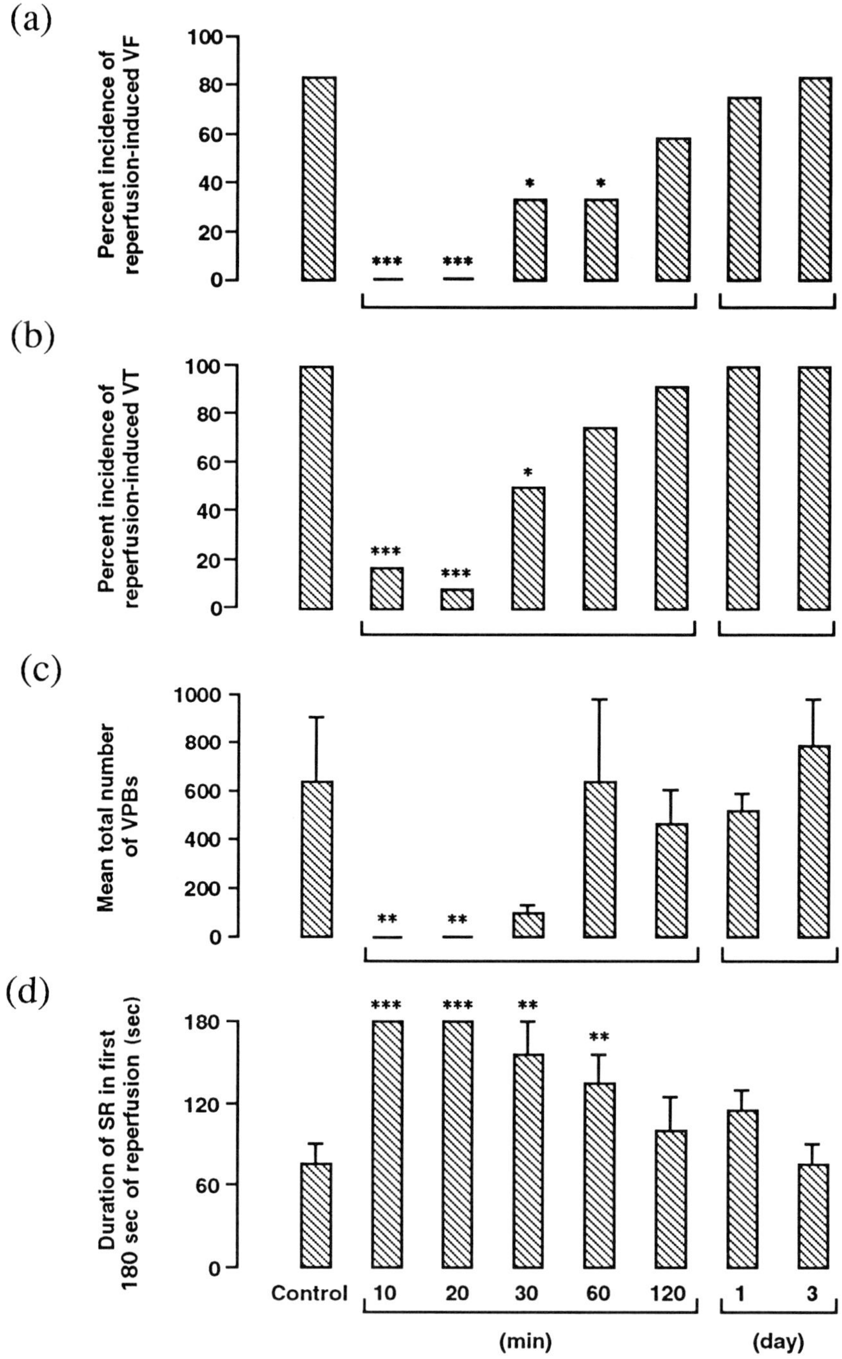
(a)
Percent incidence of reperfusion-induced VF
100
80
60
40
20
0
***
***
*
*
(b)
Percent incidence of reperfusion-induced VT
100
80
60
40
20
0
***
***
*
(c)
Mean total number of VPBs
1000
800
600
400
200
0
**
**
(d)
Duration of SR in first 180 sec of reperfusion (sec)
180
120
60
0
***
***
**
**
Control
10
20
30
60
120
1
3
(min)
(day)

Any protective myocardial adaptation induced by ischaemic preconditioning could affect this bell-shaped relationship in three distinct ways (*Figure 2*). First, since ischaemic preconditioning delays the development of ischaemia-induced myocardial necrosis (Murry *et al.*, 1986) it could increase tolerance to factors generating arrhythmias in a similar way, effectively delaying their onset. This would be expected to result in a shift in the bell-shaped curve to the right (*Figure 2a*). A second possibility is that, conversely, serial ischaemic episodes may lead to cumulative ischaemic damage. This would be expected to shift the bell-shaped curve to the left. However, this could still lead to a reduction in severity of reperfusion-induced arrhythmias being recorded, dependent on the duration of ischaemia studied (*Figure 2b*). The third possibility is that preconditioning ischaemia might act to reduce the incidence of reperfusion-induced arrhythmias following all ischaemic durations without altering the ischaemic duration associated with maximum severity of arrhythmias (i.e. the curve would not be shifted right or left but would be reduced in amplitude; *Figure 2c*).

It is clear from *Figure 2* that if only a single ischaemic duration is studied, it is possible for the most careful study of preconditioning against arrhythmias to be misleading, since a reduction in arrhythmia incidence could be observed without the tolerance of the myocardium to ischaemia *necessarily* having been increased. Which one of these three scenarios occurs in practice has now been investigated in a blood-perfused isolated rat heart preparation (Lawson *et al.*, 1993a,b). Preconditioning was accomplished using three cycles of 5 min of regional ischaemia and 5 min of reperfusion, and reperfusion-induced arrhythmias were analysed following ischaemic durations ranging from 5 to 40 min. For each ischaemic duration studied, a reduction in the severity of reperfusion-induced arrhythmias was seen. Peak incidences of reperfusion-induced VT and VF occurred after 10 min ischaemia in

---

Figure 1. Relationship between duration of 'recovery' and vulnerability of heart to arrhythmias induced by reperfusion after a second episode (5 min) of regional ischaemia *in vivo*. Hearts (n = 12 per group) were subjected to 5 min regional ischaemia followed by reperfusion for 10, 20, 30, 60, 120 min, 1 or 3 days, after which regional ischaemia was again induced for a period of 5 min. Histograms show the incidence of ventricular fibrillation (a), incidence of ventricular tachycardia (b), mean total number of ventricular premature complexes (c), and mean time in normal sinus rhythm during the first 3 min of reperfusion (d). Results are compared with control group, which corresponds to value obtained after first period of ischaemia and reperfusion. * $p<0.05$, ** $p<0.01$, *** $p<0.0001$ vs. control group. Redrawn using data from Shiki and Hearse (1987).

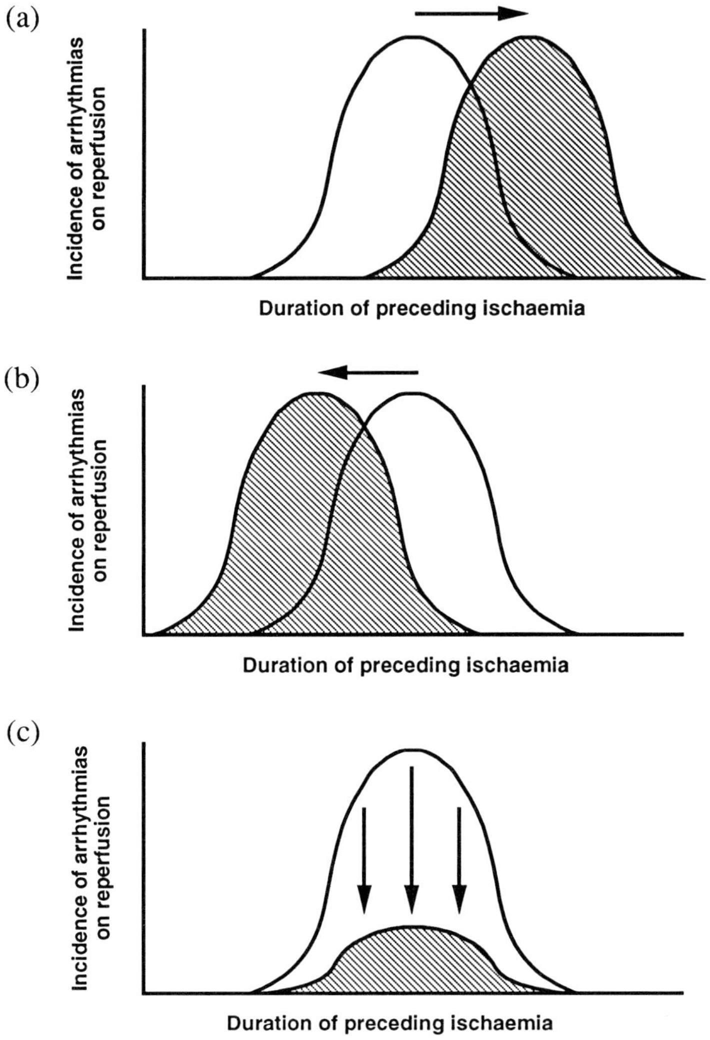

Figure 2. Schematic diagram of the possible effects of ischaemic preconditioning on the 'bell-shaped' relationship between the incidence of reperfusion-induced arrhythmias and the duration of preceding ischaemia. (a) If preconditioning acts by increasing ischaemic tolerance a shift in the relationship to the right might be expected (arrow); dependent on the duration of preceding ischaemia either a reduction or an increase in the severity of reperfusion-induced arrhythmias could be recorded. (b) If cumulative ischaemic damage occurred, a shift in the relationship to the left might be expected (arrow); again, dependent on the duration of preceding ischaemia, either a reduction or an increase in the severity of reperfusion-induced arrhythmias could be recorded. (c) If preconditioning has a truly anti-arrhythmic effect distinct from its anti-ischaemic actions this might be expected to result in a reduction in arrhythmia severity irrespective of the duration of preceding ischaemia studied (arrows).

both control and preconditioned hearts. This showed that no temporal shift had occurred in the bell-shaped relationship but the amplitude of the curve was reduced (*Figure 3*). This suggested that whatever myocardial adaptation is induced by preconditioning is not primarily an increase in ischaemic tolerance but an additional and distinct anti-arrhythmic action.

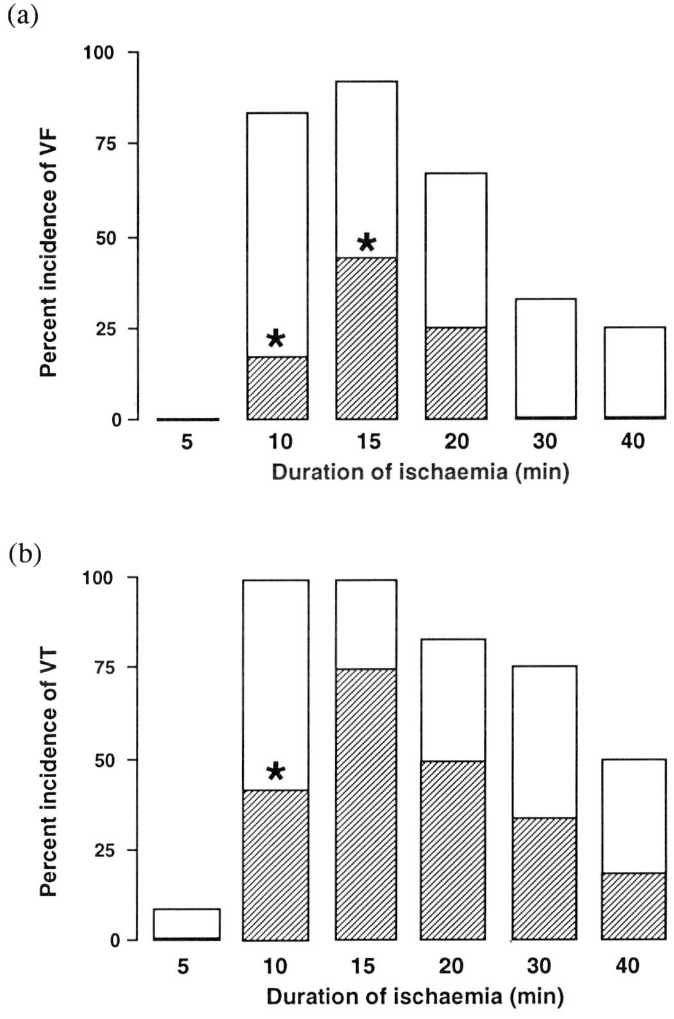

Figure 3. Effect of preconditioning on the 'bell-shaped' relationship between the incidence of reperfusion-induced arrhythmias and the duration of preceding ischaemia. The incidence of reperfusion-induced ventricular fibrillation (a) and ventricular tachycardia (b) are compared in control and preconditioned rat hearts following 5, 10, 15, 20, 30 or 40 min of ischaemia ($n = 12$ per group). Open bars = control; hatched bars = preconditioned. VF, ventricular fibrillation; VT, ventricular tachycardia. * = $p<0.05$ vs. respective control group.

2.3. Preconditioning against ischaemia-induced arrhythmias

The first studies designed specifically to assess the effect of preconditioning on arrhythmias induced by ischaemia were reported in a canine model by Vegh and colleagues (1990, 1991). When given two sequential 5 min episodes of preconditioning ischaemia, fewer than 15% of animals suffered sustained tachyarrhythmias as a consequence of preconditioning. The effect on the incidence of arrhythmias during a subsequent prolonged ischaemic episode, however, was profound. VF was abolished, the incidence of VT reduced from over 80% to less than 40%, and the mean number of VPBs reduced to 21% of the control level.

As with arrhythmias induced by reperfusion, a potential ambiguity exists as to whether protection against ischaemia-induced arrhythmias is a consequence of a true reduction in arrhythmia severity, or is simply a delaying effect such that the arrhythmias are not manifest during the ischaemic period studied. Since preconditioning delays ischaemia-induced myocardial injury it might be expected that a delaying action on ischaemia-induced arrhythmias would be seen.

2.4. Increase in ischaemic tolerance or true reduction in arrhythmia severity?

To distinguish between an absolute anti-arrhythmic action and a delaying effect, it is essential that the ischaemic duration should be long enough to ensure that delayed arrhythmias are not missed. Vegh and colleagues' first study (Vegh *et al.*, 1990) used an ischaemic duration of only 25 min. This proved insufficiently long to allow this distinction to be made. However, these authors have subsequently confirmed, in a small number of dogs, that even when the ischaemic period is extended to 60 min there is no evidence of a delaying effect on VPBs (Vegh *et al.*, 1991).

A similar effect is seen in rat hearts. Using an ischaemic period of 90 min *in vivo* Li and colleagues (1992) have shown that the reduction in arrhythmia severity is a consequence of reduced arrhythmia incidence with no evidence of any delayed arrhythmias. However, these *in vivo* studies suffer from significant data censoring owing to the number of animals dying during the course of experiments, as a consequence not only of arrhythmias but also hypotension. In one such study, for example, of 86 rats studied, 50 failed to complete the experimental protocol (Li *et al.*, 1992). This poses the question as to whether delayed arrhythmias would have occurred had the animal survived. In rats an additional complication exists, in that the anti-arrhythmic effect of preconditioning is very marked. Very few malignant arrhythmias occur during ischaemia in adequately preconditioned hearts, thus rendering a detailed analysis of their time-course impossible.

The question of whether preconditioning abolishes or merely delays the development of ischaemia-induced arrhythmias has now been addressed directly in an isolated rat heart preparation (Lawson *et al.*, 1993a,b). A distinct advantage of this model is that the hearts are isolated from the haemodynamic consequences of malignant arrhythmias, so data censoring late in the protocol is limited. The results of this study are represented in *Figure 4*. During each 5 min time band of a 40 min ischaemic period, preconditioning led to reductions in the incidences of sustained tachyarrhythmias (i.e. VF and VT) and VPBs. It was also found that increasing the number of preconditioning cycles induced additive anti-arrhythmic protection: thus, the use of two or three cycles of preconditioning ischaemia led to the virtual abolition of ischaemia-induced arrhythmias and, with a single cycle, an intermediate degree of protection occurred. However, in all groups, the time of maximal vulnerability to arrhythmias was between 10 and 20 min and few arrhythmias were seen after ischaemic durations of 25 min or more. Thus, substantial anti-arrhythmic protection was demonstrated but again there was no evidence of any significant shift in the pattern of vulnerability. This confirmed that ischaemia-induced arrhythmias are indeed abolished rather than delayed. As with reperfusion-induced arrhythmias, this argues against the adaptation induced by preconditioning being merely an increase in ischaemic tolerance.

2.5. 'Dose'-dependency of anti-arrhythmic protection by preconditioning

The induction of a graded anti-arrhythmic protection with increasing numbers of preconditioning cycles has been-termed 'dose'-dependency. This observation is borne out by other studies. While Vegh's group (Vegh *et al.*, 1992a) saw no reduction in the incidence of either VT or VF after a single 3 min preconditioning cycle in rat hearts *in vivo*, Li and colleagues were able to reduce the incidences of these arrhythmias in a similar model with three cycles of 3 min of preconditioning (Li *et al.*, 1992). The reduction in incidence of malignant arrhythmias following three preconditioning cycles in this study, in contrast to the less profound attenuation of their duration following a single cycle (Vegh *et al.*, 1992a)

(a)

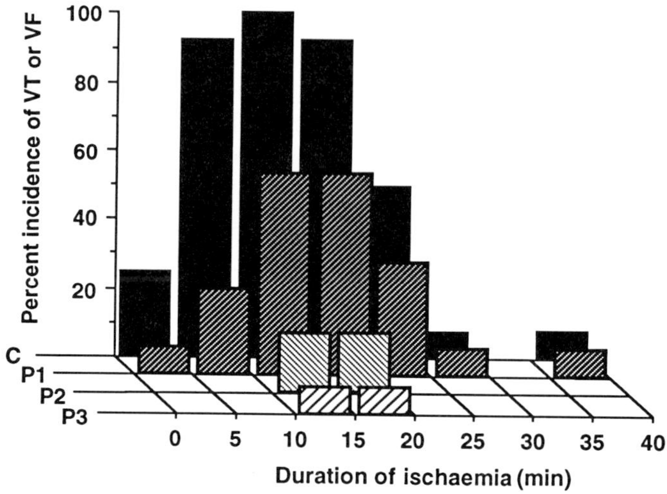

(b)

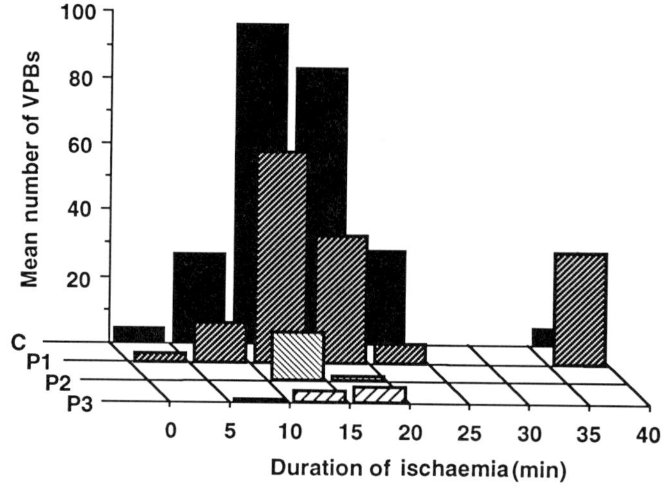

Figure 4. The effect of ischaemic preconditioning on: (a) the incidence of ventricular tachycardia or fibrillation and (b) the mean number of ventricular premature beats in isolated blood-perfused rat hearts assessed over sequential 5 min time periods during a 40 min ischaemic episode. VF, ventricular fibrillation; VT, ventricular tachycardia; VPB, ventricular premature beats. Preconditioning induced by cycles of 5 min of ischaemia and 5 min of reperfusion. Study groups (n = 12 per group): C, control, P1, one preconditioning cycle, P2, two preconditioning cycles and P3 , three preconditioning cycles.

in a virtually identical model, might indicate the occurrence of additional anti-arrhythmic benefit from the second and third preconditioning cycles.

As mentioned above, this pattern of 'dose'-dependent protection with up to three preconditioning cycles also occurs for reperfusion-induced arrhythmias in this model (Lawson *et al.,* 1993a,b). Intriguingly, it appears that the conditions required for inducing a preconditioned state are different for necrosis and arrhythmias. Downey has reported (personal communication) that in rat hearts *in vivo* a single cycle of 5 min of ischaemia is sufficient virtually to abolish ischaemia-induced arrhythmias but that three cycles are required before protection against myocardial necrosis can be demonstrated.

2.6. Species differences

A remarkable property of ischaemic preconditioning has been the consistency of protection against myocardial necrosis reported by different laboratories. Indeed, every species studied to date has shown a reduction in infarct size following preconditioning (Lawson and Downey, 1993; Przyklenk and Kloner, 1995). The situation is rather different when arrhythmias are considered. It is well established that different species differ significantly in the ease with which ventricular arrhythmias can be induced by ischaemia or by reperfusion. It is, therefore, hardly surprising that protection afforded by preconditioning against arrhythmias occurs less universally than against myocardial necrosis. In rats, protection has consistently been observed, by ourselves and other groups (Hagar *et al.,* 1991; Lawson *et al.,* 1993a,b; Li *et al.,* 1992; Shiki and Hearse, 1987; Vegh *et al.,* 1992a). In dogs, some authors have shown profound anti-arrhythmic protection (Vegh *et al.,* 1990, 1992a,b), some have not observed this (Murry *et al.,* 1986), and one report observed an increase in the severity of arrhythmias with preconditioning (Reimer *et al.,* 1990). To date there has been only a single report of anti-arrhythmic activity in pigs (Wainwright and Parratt, 1990), and none in rabbits. It is not clear whether these species differences reflect gross differences, for example in coronary artery anatomy or fundamental differences in signalling at the molecular level.

3. Molecular events involved in preconditioning against arrhythmias

3.1. Myocardial adaptations subserving protection against arrhythmias

The striking similarities between anti-arrhythmia and anti-infarct protection by preconditioning make it reasonable to suggest that the molecular processes involved overlap to a large degree. Ultimately, however, myocardium which has acquired a resistance to arrhythmias must have had better preservation of ionic homeostasis than unprotected myocardium, whereas the essential requirements for protection against necrosis may be quite different. From many of the studies outlined above it is clear that in certain protocols it is perfectly possible to protect against necrosis without the prevention of arrhythmias and vice versa. Indeed, it may be that the anti-arrhythmic effect of preconditioning is a serendipitous epiphenomenon resulting from whatever cellular adaptations induced by the stimulus used to induce the 'anti-ischaemic' effect of preconditioning. This possibility remains to be formally tested. The molecular mechanisms underlying separate protective effects on different endpoints may well therefore, be very different.

Protection by endogenous substrates. The demonstration that preconditioning by ischaemia can be mimicked pharmacologically has prompted the search for endogenous molecules which take part in the process. Most of these studies have been concerned with infarct size as an endpoint. Given the inherent species variation in susceptibility to arrhythmias, it is perhaps not surprising that the few studies that have been performed specifically to identify the molecular mechanisms underlying the anti-arrhythmic protection have sometimes yielded apparently conflicting results.

Adenosine: although a wealth of evidence exists to suggest an important, and possibly pivotal, role for adenosine in mediating protection against myocardial necrosis in several species (Mullane, 1992) there is none to support such a role in protection against arrhythmias. This may be due to the majority of arrhythmia studies having been performed in the rat, a species in which adenosine does not appear to be involved in the mechanism of preconditioning (Liu and Downey, 1992).

Prostacyclin: in anaesthetized dogs, Vegh and colleagues (1990) have shown that they could block the anti-arrhythmic protection induced by preconditioning by the administration of the cyclo-oxygenase inhibitor meclofenamate. They suggested that, during preconditioning, ischaemic myocardium produces a prostaglandin analogue, presumably prostacyclin, which is known to be capable of reducing the severity of ischaemia-induced arrhythmias (Coker and Parratt, 1983). However, in the rat, Li and Kloner (1992) found that aspirin failed to prevent protection against either necrosis or ischaemia-induced arrhythmias. Similarly, Liu and colleagues (1992) have shown that neither meclofenamate nor aspirin prevent protection against infarction in rabbits. These apparently conflicting results might be explained by differences in species, differences in the protocols employed or the pharmacological agents used, or the possibility of a direct pro-arrhythmic effect of meclofenamate.

Nitric oxide: two lines of evidence point to a role for nitric oxide (NO) in the anti-arrhythmic effect of preconditioning. Blocking the effects of NO by inhibiting guanylate cyclase with intracoronary methylene blue reversed the relatively weak anti-arrhythmic protection of preconditioning in a canine heart (Vegh *et al.*, 1992c). The same authors have shown that inhibiting NO production with L-NAME (N^G-nitro-L-arginine methyl ester) reduces the incidence of VF during reperfusion (Vegh *et al.*, 1992b). They speculated that ischaemia causes local bradykinin release from endothelial cells which releases prostacyclin and subsequently stimulates local NO production. These results concur with a report in which L-NAME was shown to reduce infarct size in an isolated rat heart (Weselcouch *et al.*, 1995). This appeared to be due to an adenosine-dependent mechanism, since the effect could be blocked by 8-*p*-sulphophenyltheophylline (SPT), a specific adenosine receptor antagonist. However, when recovery of contractile function was used as an endpoint in isolated rat hearts, no effect on ischaemic preconditioning was found by prior treatment with L-NAME (Woolfson *et al.*, 1995). Again, it may be that these differences represent true differences between species, or differential effects on the different endpoints selected in each study. It is not currently possible to determine what contribution NO might be making at the molecular level in the effect of preconditioning on arrhythmias.

Ionic homeostasis. Although the generation of arrhythmias depends ultimately on some perturbation of ionic homeostasis in the myocardium, relatively few studies have concentrated on investigating trans-sarcolemmal ionic fluxes during the preconditioning stimulus and the subsequent ischaemic period.

Potassium: the arrhythmogenic potential of the regional hyperkalaemia seen during ischaemia is well recognized. However, potassium accumulation alone is not sufficient to precondition against arrhythmias, since repeated episodes of hyperkalaemic perfusion elicit repeated arrhythmic episodes (Curtis, 1991). Direct measurement of extracellular potassium with potassium-sensitive microelectrodes indicates that with repeated coronary artery occlusions, the later occlusions lead to a less rapid, but ultimately more profound, elevation of myocardial extracellular potassium (Fleet *et al.*, 1985). It is conceivable that this could have a local cardioplegia-like effect and protect against necrosis.

Further evidence that potassium homeostasis may be important in preconditioning comes from studies implicating the sarcolemmal adenosine triphosphate (ATP)-sensitive potassium (K_{ATP}) channel in the mechanism of preconditioning in many species. These have been reviewed recently (Parratt and Kane, 1994). However, the effect of drugs, such as pinacidil or bimakalim, which open these channels, is complex and controversial (Hearse 1995). Some studies show K_{ATP} channel activation to be pro-arrhythmic, others show it to be anti-arrhythmic, still others show no effect. Channel inhibition by glibenclamide can prevent protection by preconditioning against myocardial necrosis in dogs (Gross and Auchampach, 1992) but this effect was not seen with preconditioning against arrhythmias (Vegh *et al.*, 1993). Overall, there is no compelling evidence that K_{ATP} activation is a powerful anti-arrhythmic intervention against either ischaemia- or reperfusion-induced arrhythmias (Hearse, 1995). As such it seems very unlikely that this channel can play a pivotal role in the anti-arrhythmic effects of preconditioning.

Calcium: increases in cytosolic calcium loading, as a consequence of ischaemia and/or reperfusion, may contribute significantly to arrhythmogenesis by a number of pro-arrhythmic mechanisms including action potential shortening, the induction of a depolarizing inward current and impaired intracellular conduction (Opie and Coetzee, 1988). Both biochemical (Tosaki *et al.*, 1994) and magnetic resonance (Steenbergen *et al.*, 1993) studies have indicated that calcium loading consequent upon ischaemia is significantly attenuated by preconditioning. Hypoxic perfusion, which seems to be as effective as ischaemia in inducing preconditioning (Shizukida *et al.*, 1992; Zhai *et al.*, 1993), has also been shown to limit the uptake of calcium (Tani and Neely, 1989). It has recently been shown that preconditioning alters the rate of sarcoplasmic reticulum (SR) calcium release during ischaemia (Zucchi *et al.*, 1995). This suggests a plausible mechanism by which cytosolic calcium overload could be attenuated. These authors also pointed out that since SR calcium channel opening probability increases with increasing cytosolic calcium concentration, a positive feedback loop exists during ischaemic conditions. If preconditioning interferes with this amplification mechanism, then even a relatively modest inhibition of SR calcium release channel function could have a profound effect on cellular excitability by limiting the amount of calcium released into the cytosol during a subsequent ischaemic period.

Protons: proton washout on reperfusion can lead to an exacerbation of ischaemia-induced sodium loading (by sodium–proton exchange) and consequent calcium loading due to sodium–calcium exchange (Lazdunski *et al.*, 1985; Tani and Neely, 1989). This mechanism may, therefore, be involved in the initiation of reperfusion-induced arrhythmias. Support for this idea has been provided by Avkiran and Ibuki (1993) who have shown that acidotic reperfusion can prevent reperfusion-induced arrhythmias by attenuating the sudden development of a proton gradient across the sarcolemmal membrane. Direct measurement of extracellular pH and measurement of intracellular pH (pH_i) by magnetic resonance spectroscopy has shown a reduction in both parameters during serial ischaemic episodes (Fleet *et al.*, 1985; Kida *et al.*, 1991). This may be secondary to a differential redox potential in preconditioned myocardium as compared to controls, since the attenuated fall in pH_i seen in preconditioned hearts during ischaemia is abolished when hearts are preconditioned in the presence of *N*-acetylcysteine, a glutathione precursor and anti-oxidant (Chen *et al.*, 1995).

Oxidant stress. The cellular mechanisms underlying reperfusion-induced arrhythmias appear to be substantially different from those responsible for arrhythmias induced by ischaemia. While oxidant stress appears to be crucial in the precipitation of reperfusion-induced arrhythmias (Bernier *et al.*, 1986; Woodward and Zakaria, 1985), there have been conflicting reports on the effect of preconditioning on tissue anti-oxidant defences (Hoshida *et al.*, 1991; Turrens *et al.*, 1992). The effect of free radical scavengers on protection against myocardial necrosis is similarly controversial (Iwamoto *et al.*, 1991; Richard *et al.*, 1993; Tanaka *et al.*, 1994) and no direct measurements of the effect of preconditioning on free radical production have been reported. Anti-arrhythmic protection induced by preconditioning has been prevented by superoxide dismutase (SOD; Osada *et al.*, 1994) but unfortunately this study did not include a group showing the effect of SOD in the non-preconditioned group of animals.

An intriguing hypothesis has recently been proposed by Chen and colleagues who have shown that alterations in myocardial thiol redox state seem to be an important signal in the molecular mechanism of preconditioning (Chen *et al.*, 1995). The addition of *N*-acetylcysteine during their preconditioning protocol blocked not only glutathione loss during preconditioning, but also the protective effect of preconditioning, as assessed by the recovery of contractile function and creatine kinase leakage on reperfusion. They suggested that thiol oxidation during preconditioning may be protective, contrary to the more generally assumed notion that preconditioning protects against oxidation. However, while it is perfectly plausible that an altered redox state should act as a cellular signal, it is unlikely that glutathione depletion *per se* is important for preconditioning, since we have shown that significant glutathione depletion does not affect the incidence of malignant arrhythmias on reperfusion (Connaughton *et al.*, 1996).

Cardiac electrophysiology. The effect of serial coronary occlusions on the electrophysiology of whole hearts has been incompletely explored. While Gulker and colleagues (1977) showed that later occlusions are associated with a higher threshold for the precipitation of VF, Fleet's group (1985) demonstrated a progressive increase in activation delay with sequential coronary occlusions which could actually predispose to the development of re-entrant circuits. In contrast, the later occlusions in the study by Fleet were associated with activation block which could equally interrupt such a circuit. Vegh's group (1991) has shown that anti-arrhythmic protection following ischaemic preconditioning induced by rapid pacing is associated with a reduction in the severity of conduction abnormalities within the ischaemic zone. Potentially, this could limit the dispersion of refractoriness within the ischaemic myocardium and reduce the probability of re-entrant circuits developing.

3.2. Molecular signal transduction

Considerable evidence now exists pointing to a crucial role for protein kinase C (PKC) in the cascade of cellular signals subserving preconditioning. Almost all of this evidence comes from pharmacological studies in which PKC catalytic function was activated directly or blocked. Ytrehus and colleagues (Ytrehus *et al.*, 1994) were the first to report the involvement of PKC in preconditioning. They successfully blocked the protective effect of prior ischaemia–reperfusion cycles on subsequent infarction by using staurosporine and polymyxin B in a rabbit heart. Neither of these agents is a highly selective PKC blocker when compared with their activity on certain other kinases, but the fact that their common action is to inhibit PKC activity suggested strongly that this was the relevant effect in this preparation. In addition, the authors directly activated PKC using either 4β-phorbol-12-myristate-13-acetate, a phorbol ester, or 1-oleyl-2-acetyl glycerol (OAG), an analogue of diacylglycerol, the native PKC activator. Both of these interventions were as effective as preconditioning by ischaemia in protecting against myocardial necrosis. A key role for PKC is supported by evidence from an *in vivo* rat study (Speechly-Dick *et al.*, 1994). In this study, chelerythrine was used as the PKC blocker and 1,2-dioctanoyl-*sn*-glycerol (DOG), another diacylglycerol analogue, as a direct PKC agonist. DOG mimicked the protection against necrosis observed with ischaemic preconditioning and chelerythrine abolished the protective effect of ischaemia. This suggests that activation of PKC is required for effective preconditioning, and that direct PKC activation can substitute for ischaemia as a successful intervention to protect against necrosis.

The proposal that PKC activation represents a key step between the preconditioning stimulus and subsequent cellular adaptation is an attractive one for many reasons. Preconditioning by ischaemia can be mimicked by many pharmacological agents binding to sarcolemmal receptors. In all systems studied so far, these receptors all couple to PKC, via intra-membrane and cytosolic intermediaries. Thus the adenosine (Downey *et al.*, 1994), α_1-adrenergic (Tsuchida *et al.*, 1994), muscarinic M_2 (Fleming *et al.*, 1992), angiotensin II (Liu *et al.*, 1995) and bradykinin (Goto *et al.*, 1995) receptors are all capable of activating PKC. Stimulation of any of them can mimic ischaemic preconditioning against myocardial necrosis, and in every case apart from the M_2 receptor such 'pharmacological' preconditioning has been shown to be PKC dependent. Earlier complementary evidence showing that intra-membrane G proteins are involved in the mechanism of preconditioning (Parratt, 1995; Thornton *et al.*, 1993) is neatly accommodated by this scheme, since G proteins are known to be involved in signal transduction between sarcolemmal receptors and PKC (Fleming *et al.*, 1992).

Regrettably, no study to date has examined the role of PKC in preconditioning using arrhythmias as a primary endpoint. Only one study (Speechly-Dick *et al.*, 1994) has considered ventricular arrhythmias as a secondary endpoint in a study of the role of PKC in preconditioning, although several others have looked at the G protein component of this signalling cascade (Hu and Nattel, 1995; Lawson *et al.*, 1993; Piacentini *et al.*, 1993, 1995). Speechly-Dick and colleagues found that while preconditioning using ischaemia abolished both VT and VF, blocking PKC function with chelerythrine did not abolish this protection. Results from the G protein studies are controversial. Piacentini's studies, and that of Hu and Nattel, found that treatment with pertussis toxin, an inhibitor of G proteins, interfered with the reduction in the frequency of ventricular arrhythmias induced by preconditioning with ischaemia. In contrast, Lawson and colleagues found no differences in the incidence of VT or VF between hearts preconditioned with or without pertussis toxin pretreatment. While these apparently conflicting results may reflect anything from the different experimental models used to a true biological difference between G proteins and PKC, they serve also to emphasize that definitive studies are required to clarify the molecular mechanisms underlying preconditioning's effect on arrhythmias.

The above descriptions all refer to what has become known as 'classical' or 'early' preconditioning. More recently it has become apparent that a 'second window of protection' opens some 24 h after the original preconditioning stimulus (for review see Yellon and Baxter, 1995) dependent, at least in part, on the synthesis of protective proteins in the intervening period. Protection during this 'second window' appears to delay the onset of myocardial necrosis, but Shiki and Hearse (1987) found no evidence of such an effect on reperfusion arrhythmias, at up to 72 h after the preconditioning stimulus. Further studies will be required to determine whether such protection can be induced against arrhythmias caused by ischaemia and are discussed by Parratt *et al.* in Chapter 5.

4. Arrhythmias – a unique endpoint in preconditioning?

Much of the evidence reviewed above suggests that significant differences exist between the protection induced by preconditioning against arrhythmias, and that induced against endpoints such as myocardial necrosis. Inevitably many comparisons remain incomplete since studies to answer the relevant questions have yet to be undertaken. Nonetheless, arrhythmias appear to show unique characteristics when compared with other endpoints used in studies of myocardial preconditioning. For example:

(i) although most models examined show remarkable consistency in protection against infarction there are considerable species and model differences in arrhythmia protection. Part of this observed difference may be simply due to true species differences in inducing arrhythmias in some models in the control condition. The rat heart for instance, is very susceptible to arrhythmias, but appears to be preconditioned by different mechanisms to species such as rabbit and dog.

(ii) Anti-infarct protection by preconditioning is capable only of delaying the onset of irreversible myocardial injury, in effect buying extra time for effective reperfusion. In contrast, preconditioning against arrhythmias appears to be capable of causing an absolute reduction in arrhythmia incidence which is not dependent on subsequent reperfusion. In this sense preconditioning is truly anti-arrhythmic.

(iii) The temporal boundaries of the envelope of protection provided by preconditioning appear to be different when arrhythmias are compared to myocardial necrosis, for example. Only 'classical' or 'early' protection has been shown in this context, with no evidence of a 'second window of protection' in the period up to 72 h following the preconditioning stimulus.

(iv) The phenomenon of 'dose'-dependent protection is clearly demonstrated for protection against both ischaemia- and reperfusion-induced arrhythmias; this is not yet the case for protection against necrosis.

(v) The relatively small amount of direct evidence suggests that the molecular mechanisms underlying preconditioning in each context are different.

5. Designing and assessing studies of preconditioning using arrhythmias as an endpoint

It should now be apparent that the design of studies focusing on the effect of preconditioning on arrhythmias will differ significantly from that concentrating on another endpoint such as infarct size. We would suggest that an ideal protocol should:

(i) use relatively short durations of preconditioning ischaemia to limit arrhythmias induced by the preconditioning cycles. Alternatively, the ischaemic duration in the second preconditioning cycle can be made longer than the first, a strategy we have employed in our laboratory.

(ii) Make a large portion of the left ventricle ischaemic to ensure a high incidence of arrhythmias in control hearts. Our studies with rat hearts normally involve proximal left coronary ligation and the occluded zone has represented over 40% of the ventricular mass. Since studies of infarct size generally involve longer experimental durations they generally use much smaller occluded zone sizes to limit the number of deaths occurring due to hypotension and cardiac failure during the longer experiments.

(iii) Use sufficiently large group sizes to allow the identification of a protective effect. In comparison with studies of infarction, larger sample sizes are required to show significant changes in arrhythmia incidence owing to differences in the statistical techniques employed. Our arrhythmia studies generally use $n = 12$ per group whereas it is unusual for the group size to exceed eight in studies of infarct size. In species such as the rabbit where, even with a large occluded zone, malignant arrhythmias are relatively difficult to induce, much larger group sizes may be required to demonstrate statistically significant benefit. Power analyses should be conducted prior to experimentation to permit the planning of adequate group sizes.

(iv) Use ischaemic durations appropriate for the species, model and arrhythmia type under study. Because of the temporal relationships of ischaemia- and reperfusion-induced arrhythmias described above, it is impossible to study both optimally using the same ischaemic duration. Data should not therefore be pooled from arrhythmias induced by both ischaemia and reperfusion. This practice is relatively common but since the mechanisms involved in ischaemia- and reperfusion-induced arrhythmias are widely different, such an approach is of dubious scientific merit.

(v) Comply with the Lambeth conventions (Walker *et al.*, 1988).

In summary, we would strongly urge that arrhythmias should be assessed in purpose-designed protocols rather than examined as a secondary endpoint. Ideally, ischaemia- and reperfusion-induced arrhythmias should also be studied using separate experimental protocols.

6. Clinical potential of preconditioning as an anti-arrhythmic intervention

While the potential for preconditioning as a therapeutic intervention in man may have been overestimated (Connaughton and Hearse, 1996; Przyklenk and Kloner, 1995), if human myocardium could be successfully preconditioned against arrhythmias this might lead to enormous clinical benefits. In the very early stages of acute myocardial infarction (AMI) mortality is believed to be more dependent on the generation of malignant ventricular arrhythmias rather than to primary pump failure. Unless immediate resuscitation is available, patients who develop such malignant arrhythmias will die suddenly. In the face of such arrhythmias, the outlook for the patient depends on the type of arrhythmia, the response time of emergency medical services and whether or not on-site resuscitation is begun by bystanders (Ritter *et al.*, 1985). Numerically, this group of people may be very large. Approximately 170 000 people die of AMI in the United Kingdom each year (UK government figures), and as many as 50% of deaths from AMI may occur in the first 2 h after the onset of symptoms.

While it is convenient experimentally to make a distinction between ischaemia- and reperfusion-induced arrhythmias, causation in clinical practice is almost certainly much more complicated. Malignant arrhythmias occurring immediately post-myocardial infarction, during unstable angina, or during acute coronary spasm may indeed be due simply to ischaemia, with or without subsequent reperfusion. However, the sudden onset of ventricular arrhythmias is thought to be the cause of sudden cardiac death in many other conditions. In these cases, the cause of the arrhythmias may not conveniently be ascribed into one or other of these two classes. Rather, the arrhythmia substrate may depend on anatomical [left ventricular (LV) outflow obstruction, LV aneurysm], mechanical (dilated cardiac chamber), ionic (disturbances of K^+, Ca^{2+} or acid–base balance), infective or inflammatory causes. These very diverse aetiologies may share a common provocative factor, such as conduction heterogeneity, which would be capable of interfering with normal conduction sufficiently to precipitate a sustained ventricular arrhythmia.

Current pharmacological anti-arrhythmic therapy is generally unsatisfactory as prophylaxis against malignant arrhythmias. In certain contexts anti-arrhythmic drugs such as amiodarone may reduce the risk of sudden death (Dolack, 1994; McKenna *et al.*, 1985), but such agents have yet to demonstrate unequivocally a beneficial effect on mortality in patients with ischaemic heart disease. Contemporary practice is therefore forced to rely more heavily on additional or separate non-pharmacological treatments for patients at risk of life-threatening arrhythmias. Available choices include permanent pacing, implantable defibrillators, the catheter ablation of anatomical substrates of arrhythmias and 'anti-arrhythmia' surgery. However, use of these devices and procedures inevitably carries some risk, they are expensive and are only available in certain centres.

When the molecular basis of adaptation in preconditioning is understood, the potential will exist to offer such patients rational therapies against arrhythmias occurring in the contexts of acute ischaemia and reperfusion. Furthermore, if preconditioning strategies were shown to protect against arrhythmias in other settings, it should even be possible to design pharmacological agents to protect against arrhythmias of more diverse aetiologies.

7. Concluding comments

Protection against arrhythmias by preconditioning is qualitatively different from that against necrosis. Most strikingly, preconditioning has a powerful anti-arrhythmic effect over and above its anti-ischaemic actions. If this anti-arrhythmic property is to be adequately assessed it is essential that studies should be designed to look specifically at its mechanisms using appropriate models and experimental protocols. The design of such protocols is more complicated than for infarct size studies. A good preconditioning protocol should precipitate as few arrhythmias as possible, and it is also essential that it should not cause spurious 'protection' as a consequence of cumulative irreversible ischaemic damage.

Several features indicate that protection against arrhythmias might be qualitatively different from that against necrosis. These include its 'dose'-dependency, its occurrence in the context of reversible myocardial injury, the lack of effect of preconditioning on the temporal pattern of arrhythmogenesis and the likely differences in the molecular mechanisms involved. Since it remains far from clear that the diverse protective effects of preconditioning are mediated by common mechanisms, it is essential when reporting results that both the preconditioning stimulus and the primary endpoint should be specified.

References

Avkiran M and Ibuki C. (1993) Reperfusion-induced arrhythmias: a role for washout of extracellular protons? *Circ. Res.* **71,** 1429–1440.

Barber MJ. (1983) Effect of time interval between repeated brief coronary artery occlusions on arrhythmia, electrical activity and myocardial blood flow. *J. Am. Coll. Cardiol.* **2,** 699–705.

Bernier M, Hearse DJ and Manning AS. (1986) Reperfusion-induced arrhythmias and oxygen-derived free radicals. *Circ. Res.* **58,** 331–340.

Chen W, Gabel S, Steenbergen C and Murphy E. (1995) A redox-based mechanism for cardioprotection induced by ischemic preconditioning in the perfused rat heart. *Circ. Res.* **77,** 424–429.

Coker SJ and Parratt JR. (1983) Prostacyclin – anti-arrhythmic or arrhythmogenic? Comparison of the effects of intravenous and intracoronary prostacyclin and ZX 36374 during coronary artery occlusion and reperfusion in the anaesthetised greyhound. *J. Cardiovasc. Pharmacol.* **5,** 557–567.

Connaughton M and Hearse DJ. (1996) Three questions about preconditioning. *Basic Res. Cardiol.,* **91:** 12–15.

Connaughton M, Kelly FJ, Haddock PS, Hearse DJ and Shattock MJ. (1996) Ventricular arrhythmias induced by ischaemia–reperfusion are unaffected by myocardial glutathione depletion. *J. Mol. Cell. Cardiol.,* in press.

Curtis MJ. (1991) The rabbit dual coronary perfusion model: a new method for assessing the pathological relevance of individual products of the ischaemic milieu: role of potassium in arrhythmogenesis. *Cardiovasc. Res.* **25,** 1010–1022.

Dolack GL. (1994) Clinical predicors of implantable cardioverter-defibrillator shocks (results of the CASCADE trial). Cardiac arrest in Seattle, conventional versus amiodarone drug evaluation. *Am. J. Cardiol.* **73,** 237–241.

Downey JM, Cohen MV, Ytrehus K and Liu Y. (1994) Cellular mechanisms in ischemic preconditioning: the role of adenosine and protein kinase C. *Ann. NY Acad. Sci.* **723,** 82–98.

Fleet WF, Johnson TA, Graebner CA and Gettes LS. (1985) Effect of brief ischemic episodes on extracellular K, pH and activation in the pig. *Circulation,* **2,** 922–932.

Fleming JW, Wisler PL and Watanabe AM. (1992) Signal transduction by G proteins in cardiac tissue. *Circulation,* **85,** 420–433.

Goto M, Liu Y, Yang X-M, Ardell JL, Cohen MV and Downey JM. (1995) Role of bradykinin in protection of ischemic preconditioning in rabbit hearts. *Circ. Res.* **77,** 611–621.

Gross GJ and Auchampach JA. (1992) Blockade of ATP-sensitive potassium channels prevents myocardial preconditioning in dogs. *Circ. Res.* **70,** 223–233.

Gulker J, Kramer B, Stephan K and Meeseman W. (1977) Changes in ventricular fibrillation threshold during repeated short-term coronary occlusion and release. *Basic Res. Cardiol.* **72,** 547–562.

Hagar JM, Hale SL and Kloner RA. (1991) Effects of preconditioning ischemia on reperfusion arrhythmias after coronary artery occlusion and reperfusion in the rat. *Circ. Res.* **68,** 61–68.

Harris AS. (1950) Delayed development of ventricular ectopic rhythms following experimental coronary occlusion. *Circulation,* **1,** 1318–1327.

Hearse DJ. (1995) Activation of ATP-sensitive potassium channels: a novel pharmacological approach to myocardial protection? *Cardiovasc. Res.* **30,** 1–17.

Hoshida S, Kuzuya T, Fuji H, Oe H, Kimura Y, Sasaki T, Kitabatake A and Tada M. (1991) Ischemic preconditioning affects free radical generating and scavenging systems in canine myocardial infarction. *Circulation,* **84** (Suppl. II), 192.

Hu K and Nattel S. (1995) Mechanisms of ischemic preconditioning in rat hearts. Involvement of a_{1B}-adrenoceptors, pertussis toxin-sensitive G proteins, and protein kinase C. *Circulation,* **92,** 2259–2265.

Iwamoto T, Miura T, Adechi T, Noto T, Ogawa T, Tsuchida A and Iimura O. (1991) Myocardial infarct size-limiting effect of ischemic preconditioning was not attenuated by oxygen free-radical scavengers in the rabbit. *Circulation,* **83,** 1015–1022.

Kida M, Fujiwara H, Ishida M, Kawai C, Ohura M, Miura I and Yabuuchi Y. (1991) Ischemic preconditioning preserves creatine phosphate and intracellular pH. *Circulation,* **84,** 2495–2503.

Lawson CS and Downey JM. (1993) Preconditioning: state of the art myocardial protection. *Cardiovasc. Res.* **27,** 542–550.

Lawson CS, Avkiran M, Shattock MJ, Coltart DJ and Hearse DJ. (1993a) Preconditioning and reperfusion arrhythmias in the isolated rat heart: true protection or temporal shift in vulnerability? *Cardiovasc. Res.* **27,** 2274–2281.

Lawson CS, Coltart DJ and Hearse DJ. (1993b) The antiarrhythmic action of ischaemic preconditioning in isolated blood perfused rat hearts does not involve functional Gi proteins. *Cardiovasc. Res.* **27,** 681–687.

Lazdunski M, Frelin C and Vigne P. (1985) The sodium/hydrogen exchange system in cardiac cells: its biological and pharmacological properties and its role in regulating internal concentrations of sodium and internal pH. *J. Mol. Cell. Cardiol.* **17,** 1029–1042.

Li Y and Kloner RA. (1992) Cardioprotective effects of ischaemic preconditioning are not mediated by prostanoids. *Cardiovasc. Res.* **26,** 226–231.

Li YW, Whittaker P and Kloner RA. (1992) The transient nature of the effect of ischemic preconditioning on myocardial infarct size and ventricular arrhythmia. *Am. Heart J.* **123,** 346–353.

Liu GS, Stanley AHS and Downey JM. (1992) Cyclooxygenase products are not involved in the protection against myocardial infarction afforded by preconditioning in rabbit. *Am. J. Cardiovasc. Path.* **4,** 56–63.

Liu Y and Downey JM. (1992) Ischemic preconditioning protects against infarction in the rat heart. *Am. J. Physiol.* **263,** H1107–1112.

Liu Y, Tsuchida A, Cohen MV and Downey JM. (1995) Pretreatment with angiotensin II activates protein kinase C and limits myocardial infarction in isolated rabbit hearts. *J. Mol. Cell. Cardiol.* **27,** 883–892.

Manning AS and Hearse DJ. (1984) Reperfusion-induced arrhythmias: mechanisms and prevention. *J. Mol. Cell. Cardiol.* **16,** 497–518.

McKenna WJ, Oakley CM, Krikler DM and Goodwin JF. (1985) Improved survival with amiodarone in patients with hypertrophic cardio-myopathy and ventricular tachycardia. *Br. Heart J.* **53,** 412–416.

Mullane K. (1992) Myocardial preconditioning. Part of the adenosine revival. *Circulation,* **85,** 845–847.

Murry CE, Jennings RB and Reimer KA. (1986) Preconditioning with ischemia: a delay of lethal cell injury in ischemic myocardium. *Circulation,* **74,** 1124–1136.

Opie LH and Coetzee WA. (1988) Role of calcium ions in reperfusion arrhythmias: relevance to pharmacological intervention. *Cardiovasc. Drugs Ther.* **2,** 623–636.

Osada M, Takeda S, Sato T, Komori S and Tamura K. (1994) The protective effect of preconditioning on reperfusion-induced arrhythmia is lost by treatment with superoxide dismutase. *Jpn. Circ. J.* **58,** 259–263.

Parratt JR. (1995) Protection of the heart by ischaemic preconditioning: mechanisms and possibilities for pharmacological exploitation. *Trends Pharmacol. Sci.* **15,** 19–25.

Parratt JR and Kane K. (1994) K_{ATP} channels in ischaemic preconditioning. *Cardiovasc. Res.* **28,** 783–787.

Piacentini L, Wainwright CL and Parratt JR. (1993) The antiarrhythmic effect of ischaemic preconditioning in isolated rat heart involves a pertussis toxin sensitive mechanism. *Cardiovasc. Res.* **27,** 674–680.

Piacentini L, Wainwright CL and Parratt JR. (1995) Effects of *Bordetella pertussis* toxin pretreatment on the antiarrythmic action of ischaemic preconditioning in anaesthetized rats. *Br. J. Pharmacol.* **114,** 755–760.

Przyklenk K and Kloner RA. (1995) Preconditioning: a balanced perspective. *Br. Heart J.* **74,** 575–577.

Reimer KA, Murry CE and Jennings RB. (1990) Cardiac adaptation to ischemia. Ischemic preconditioning increases myocardial tolerance to subsequent ischemic episodes. *Circulation,* **82,** 2266–2268.

Richard V, Tron C and Thuillez C. (1993) Ischemic preconditioning is not mediated by oxygen derived free radicals in rats. *Cardiovasc. Res.* **27,** 2016–2021.

Ritter G, Wolfe RA, Goldstein S, Landis JR, Vasu CM, Acheson A, Leighton R and Medendrop SV. (1985) The effect of bystander CPR on survival of out-of-hospital cardiac arrest victims. *Am. Heart J.* **110,** 932–937.

Shiki K and Hearse DJ. (1987) Preconditioning of ischemic myocardium: reperfusion-induced arrhythmias. *Am. J. Physiol.* **253,** H1470–H1476.

Shizukida Y, Mallet RT, Lee S-C and Downey HF. (1992) Hypoxic preconditioning of ischaemic canine myocardium. *Cardiovasc. Res.* **26,** 534–542.

Speechly-Dick ME, Mocanu MM and Yellon DM. (1994) Protein kinase C. Its role in ischemic preconditioning in the rat. *Circ. Res.* **75,** 586–590.

Steenbergen C, Perlman ME, London RE and Murphy E. (1993) Mechanism of preconditioning: ionic alterations. *Circ. Res.* **72,** 112–125.

Tanaka M, Fujiwara H, Yamasaki K and Sasayama S. (1994) Superoxide dismutase and *N*-2-mercaptopropionyl glycine attenuate infarct size limitation effect of ischaemic preconditioning in the rabbit. *Cardiovasc. Res.* **28,** 980–986.

Tani M and Neely JR. (1989) Role of intracellular Na^+ in Ca^{2+} overload and depressed recovery of ventricular function of reperfused ischemic rat hearts. *Circ. Res.* **65,** 1045–1056.

Thornton JD, Liu GS and Downey JM. (1993) Pretreatment with pertussis toxin blocks the protective effects of preconditioning: evidence for a G protein mechanism. *J. Mol. Cell. Cardiol.* **25,** 311–320.

Tosaki A, Cordis GA, Szerdalhelyi P, Engelman RM and Das DK. (1994) Effects of preconditioning on reperfusion arrhythmias, myocardial functions, formation of free radicals, and ion shifts in isolated ischemic/reperfused rat hearts. *J. Cardiovasc. Pharmacol.* **23,** 365–373.

Tsuchida A, Liu Y, Liu GS, Cohen MV and Downey JM. (1994) α_1-Adrenergic agonists precondition rabbit ischemic myocardium independent of adenosine by direct activation of protein kinase C. *Circ. Res.* **75,** 576–585.

Turrens JF, Thornton J, Barnard ML, Snyder S, Liu G and Downey JM. (1992) Protection from reperfusion injury by preconditioning hearts does not involve increased antioxidant defenses. *Am. J. Physiol.* **262,** H585–H589.

Vegh A, Szekeres L and Parratt JR. (1990) Protective effects of preconditioning of the ischemic myocardium involve cyclo-oxygenase products. *Cardiovasc. Res.* **24,** 1020–1023.

Vegh A, Szekeres L and Parratt JR. (1991) Transient ischaemia induced by rapid pacing results in myocardial preconditioning. *Cardiovasc. Res.* **25,** 1051–1053.

Vegh A, Komoro S, Szekeres L and Parratt JR. (1992a) Antiarrhythmic effects of preconditioning in anaesthetised dogs and rats. *Cardiovasc. Res.* **26,** 487–495.

Vegh A, Szekeres L and Parratt JR. (1992b) Preconditioning of the ischaemic myocardium: involvement of the L-arginine nitric oxide pathway. *Br. J. Pharmacol.* **107,** 648–625.

Vegh A, Papp JG, Szekeres L and Parratt JR. (1992c) The local intracoronary administration of methylene blue prevents the pronounced antiarrhythmic effect of ischaemic preconditioning. *Br. J. Pharmacol.* **107,** 910–911.

Vegh A, Papp JG, Szekeres L and Parratt JR. (1993) Are ATP sensitive potassium channels involved in the pronounced antiarrhythmic effects of preconditioning? *Cardiovasc. Res.* **27,** 638–643.

Wainwright CL and Parratt JR. (1990) Electrocardiographic and haemodynamic effects of myocardial preconditioning in pigs. *J. Mol. Cell. Cardiol.* **22** (Suppl. III), S77.

Walker MJA, Curtis MJ, Hearse DJ *et al.* (1988) The Lambeth Conventions: guidelines for the study of arrhythmias in ischaemia, infarction and reperfusion. *Cardiovasc. Res.* **22,** 447–455.

Weselcouch EO, Baird AJ, Sleph P and Grover GJ. (1995) Inhibition of nitric oxide synthesis does not affect ischemic preconditioning in isolated perfused rat hearts. *Am. J. Physiol.* **268,** H242–H249.

Woodward B and Zakaria MNM. (1985) Effect of some free radical scavengers on reperfusion induced arrhythmias in the isolated rat heart. *J. Mol. Cell. Cardiol.* **17,** 485–493.

Woolfson RG, Patel VC, Neild GH and Yellon DM. (1995) Inhibition of nitric oxide synthesis reduces infarct size by an adenosine-dependent mechanism. *Circulation,* **91,** 1545–1551.

Yellon DM and Baxter GF. (1995) A "second window of protection" or delayed preconditioning phenomenon: future horizons for myocardial protection? *J. Mol. Cell. Cardiol.* **27,** 1023-1034.

Ytrehus K, Liu Y and Downey JM. (1994) Preconditioning protects ischemic rabbit heart by protein kinase C activation. *Am. J. Physiol.* **266,** H1145–H1152.

Zhai X, Lawson CS, Cave AC and Hearse DJ. (1993) Preconditioning and post-ischemic contractile dysfunction: the role of impaired oxygen delivery *vs.* extracellular metabolite accumulation. *J. Mol. Cell. Cardiol.* **25,** 847–857.

Zucchi R, Ronca-Testoni S, Yu G, Galbani P, Ronca G and Mariani M. (1995) Postischemic changes in cardiac sarcoplasmic reticulum Ca^{2+} channels. A possible mechanism of ischemic preconditioning. *Circ. Res.* **76,** 1049–1056.

5

Suppression of life-threatening ventricular arrhythmias by brief periods of ischaemia and by cardiac pacing with particular reference to delayed myocardial protection

James R. Parratt[1], Agnes Vegh[2], Karoly Kaszala[2] and Julius Gy Papp[2]
[1]Department of Physiology and Pharmacology, University of Strathclyde, Glasgow G1 1XW, UK
[2]Department of Pharmacology, Albert Szent-Gyorgyi Medical University, Szeged, Hungary

1. Introduction

Ventricular tachyarrhythmias are the main factor responsible for sudden cardiac death, which remains one of the most serious problems facing modern cardiology. Although this is a multifactorial problem, left

Ischaemia: preconditioning and adaptation, edited by M.S. Marber and D.M. Yellon

ventricular dysfunction, electrical instability and particularly myocardial ischaemia are important predisposing factors for sudden cardiac death with important contributions from electrolyte balance, the autonomic nervous system and drug therapy.

2. The historical background of ventricular arrhythmias in response to brief episodes of ischaemia

It may seem surprising that, in view of the importance of ischaemia as a triggering mechanism for arrhythmogenesis, there are many examples of brief periods of ischaemia protecting the myocardium against life-threatening arrhythmias resulting from a subsequent more prolonged and severe period of ischaemia. There were, in fact, several examples of this phenomenon before the preconditioning era. One, well known to pharmacologists studying the effects of drugs on those arrhythmias that occur hours or days after an experimental myocardial infarction, is that ventricular fibrillation is less likely to occur if a coronary artery is occluded in two stages (a partial occlusion, greatly restricting flow, followed by a complete occlusion) than if the coronary artery is occluded suddenly (Harris, 1950).

There are other examples. Twenty or so years ago there was considerable interest in the use of drugs to reduce various manifestations of ischaemic injury including infarct size (e.g. Maroko *et al.*, 1971). One relatively simple way of assessing whether a drug reduces myocardial ischaemic injury is to use the technique of epicardial electrocardiographic ST-segment mapping. The usual protocol is to occlude the coronary artery, invariably in dogs, for a period of 20 min or so, release the occlusion, administer the drug and re-occlude for a similar period (e.g. Marshall and Parratt, 1977, 1979). It is well recognised that the initial occlusion should be ignored because the ST-segment changes are more marked than during subsequent coronary artery occlusions. A practical problem with this method is that ventricular arrhythmias are common during the first occlusion and that ventricular fibrillation is not uncommon. In our own studies, and in order to avoid serious ventricular arrhythmias during that first, usually 20 min, occlusion, we often occluded the artery for a short period (2–3 min) before the subsequent more prolonged (20 min) occlusions. This appeared to reduce the incidence of serious ventricular arrhythmias during the subsequent experimental procedures.

Others have investigated this phenomenon in more detail. Thus Gulker and his colleagues (1977) examined the effect of multiple coronary occlusions on the ventricular fibrillation threshold. The decrease in ventricular fibrillation threshold became increasingly less marked, and the duration increasingly shorter, with repeated coronary occlusions until a time when there was no significant decrease in fibrillation threshold at all. Another good example is the study of Barber (1983). His aim was to examine whether, in anaesthetized dogs, short (5 min) serial occlusions of the left anterior descending coronary artery produced consistent changes in myocardial blood flow, ST-segment elevation, various other electrocardiographic variables and in the incidence of arrhythmias, if these occlusion periods were separated by either short (3 min) or long (greater than 40 min) periods of reperfusion. Barber found that the number of ventricular premature complexes during a second 5 min period of occlusion was significantly less than during the first period, provided that the recovery time was 3 min, but not if it was 40 min. A similar phenomenon, this time for reperfusion arrhythmias following the release of brief periods of coronary artery occlusion, was demonstrated by Shiki and Hearse (1987). In this particular study they examined, in great detail, the effect of varying the time between two coronary artery occlusions of the same (short) duration by greatly increasing the recovery period to hours and even days. Although this was probably the earliest attempt to examine whether short coronary artery occlusions are able to protect the myocardium long after that initial stimulus, their experiments provided no evidence for the delayed myocardial protection which has since aroused considerable interest.

3. Anti-arrhythmic effects of ischaemic preconditioning

If we define ischaemic preconditioning as Murry and his colleagues (1986) did in their original paper (short periods of myocardial ischaemia increase myocardial tolerance to a subsequent more severe and prolonged period of ischaemia) then the first example of the anti-arrhythmic effect of ischaemic preconditioning was provided by Komori and his colleagues (Komori *et al.*, 1990a). They examined, in an anaesthetized rat model, whether brief periods of ischaemia reduced the severity of arrhythmias during a subsequent, more prolonged period of coronary artery occlusion. Komori, working in the Glasgow department, found that the protective effects of a single brief coronary artery occlusion were marked. A preliminary account of these studies was given to a meeting of the Physiological Society (Komori *et al.*, 1990a) and subsequently published in Japanese (Komori *et al.*, 1990b). Using an anaesthetized dog model Vegh *et al.* (1990) showed that if a prolonged coronary artery occlusion (25 min) was preceded by one or two short (5 min) coronary artery occlusions then the severity of these arrhythmias was markedly reduced. This is illustrated in *Figure 1*. When two 5 min coronary artery occlusions were used there was even an increased survival from the combined ischaemia–reperfusion insult. The most striking feature of this protection was the absence of ventricular fibrillation. The anti-arrhythmic effects of ischaemic preconditioning thus demonstrated in two species (rats, dogs), were *marked* (absence of ventricular fibrillation), *real* (if the occlusion period was extended to 60

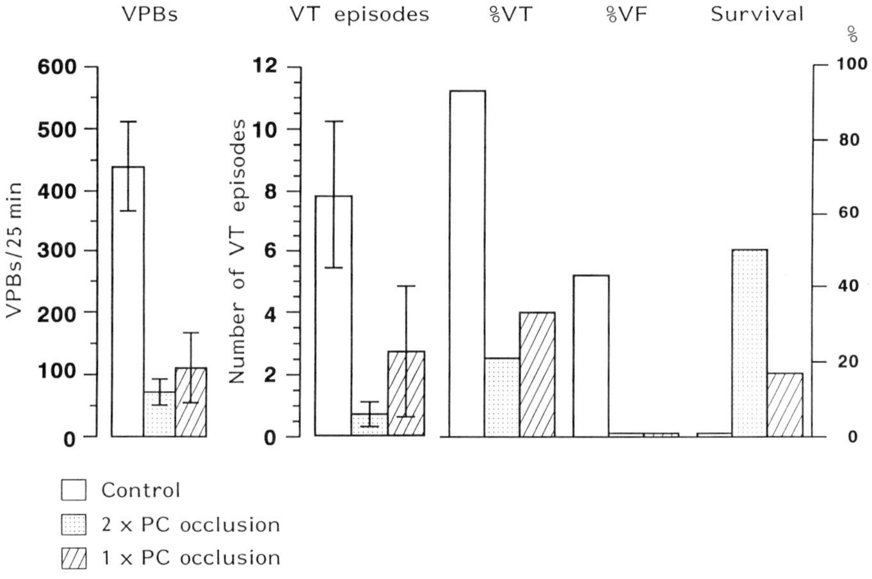

Figure 1. Ventricular arrhythmias [number of ventricular premature beats (VPB), number of episodes of ventricular tachycardia, the incidences of ventricular tachycardia (VT) and ventricular fibrillation (VF)] during a 25 min period of coronary artery occlusion in anaesthetized open-chest dogs when this occlusion was preceded by either one or two brief (5 min) occlusions of the same anterior descending branch of the left coronary artery 20 min previously. Also shown is the survival from the combined 25 min period of ischaemia followed by rapid reperfusion. One or two preconditioning coronary (PC) artery occlusions markedly reduce the severity of arrhythmias which occurred during a subsequent, prolonged period of ischaemia and reperfusion.

min the arrhythmias were not shifted to a later time during that occlusion period; Vegh *et al.*, 1992a), but *transient* (the protection was lost if the interval between the preconditioning stimulus and the prolonged occlusion was extended to 1 h rather than 20 min).

There has recently been some debate as to whether ischaemic preconditioning really does suppress ventricular arrhythmias. Although this is not the place to examine this in detail, some general observations need to be made.

There appears to be no doubt that ischaemic preconditioning reduces the severity of arrhythmias (number of ventricular premature beats, incidence and duration of ventricular tachycardia, incidence of ventricular fibrillation) in rat hearts subjected to coronary artery occlusion both *in vivo* (Komori *et al.*, 1990a,b; Li and Kloner, 1992; Vegh *et al.*, 1992a) and *in vitro* (Lawson *et al.*, 1993 a,b; Piacentini *et al.*, 1993, and illustrated in *Figure* 2).

Most studies using a canine model of preconditioning have shown a similar reduction in arrhythmia severity (e.g. Li *et al.*, 1990; Vegh *et al.*, 1990, 1992a, and reviewed by Parratt and Vegh, 1994). Others (Murry *et al.*, 1986; Przyklenk and Kloner, 1994, 1995; Vander Heide *et al.*, 1995) have been unable to confirm this marked effect. There are a number of possible explanations for this discrepancy. First, both the Jennings and Kloner groups use a barbiturate anaesthetic and provide no indication about the body (or heart) weight of the animals they have used: an important contributory factor to arrhythmia severity (Vegh *et al.*, 1992b). Barbiturates are markedly cardiodepressant and ventricular arrhythmias are not pronounced when such an anaesthetic is used; most of the early work on infarct size and on the measurement of blood flow in the ischaemic myocardium used barbiturate anaesthesia and few of them mention the confounding difficulty of serious ventricular arrhythmias in their studies. If the aim is to look at infarct size without losing too many animals through ventricular fibrillation during the early phase of myocardial ischaemia then a barbiturate is certainly the anaesthetic to use. A good example is the recent editorial by Przyklenk and Kloner in the *British Heart Journal* (Przyklenk and Kloner, 1995). *Figure 3* of that editorial shows that the total number of ventricular premature beats during a 30 min occlusion was only 23 ± 5. This compares to in excess of 400 when chloralose–urethane anaesthesia is used (e.g. Vegh *et al.*, 1992a). There is no indication in this chapter as to the time between the end of the preconditioning stimulus and the control occlusion, no indication as to which coronary artery was occluded (or where), nor whether the

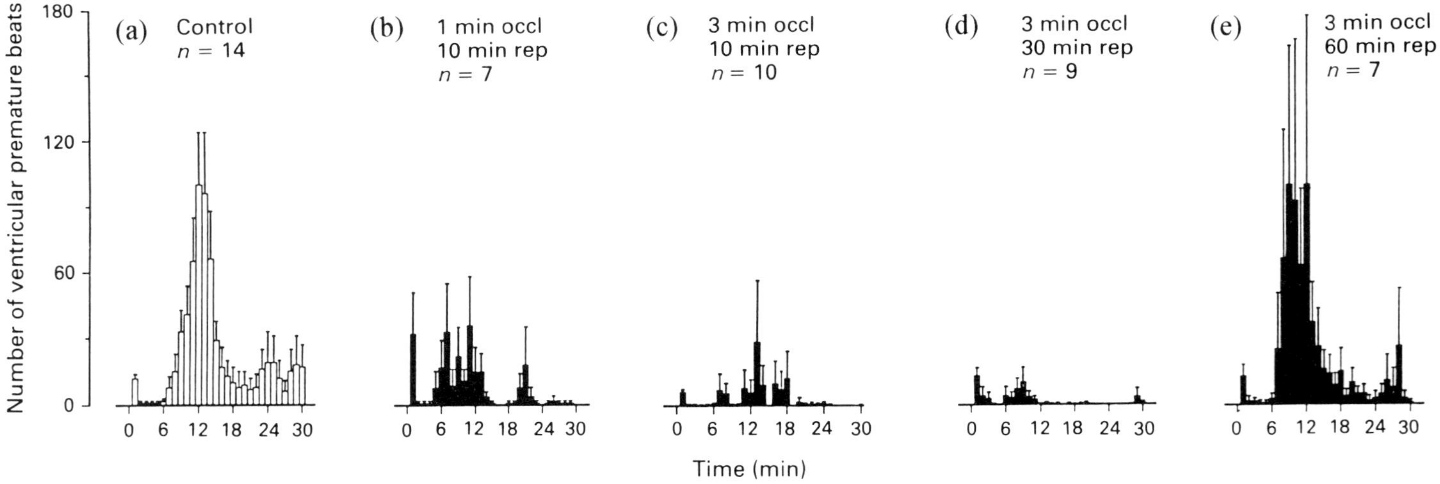

Figure 2. Distribution of ventricular premature beats in rat isolated hearts subjected to 30 min of left coronary artery occlusion. In control hearts (a) and in hearts in which this occlusion was preceded by a 1 min preconditioning occlusion followed by 10 min of reperfusion (b), or a 3 min occlusion followed by 10 (c), 30 (d) or 60 (e) min reperfusion. Either a 1 or 3 min preconditioning occlusion is effective in reducing ventricular ectopic activity but this protection is lost if the interval between preconditioning and the control occlusion is increased to 60 min (i.e. the protection is transient). Reproduced with permission from Piacentini *et al.* (1993) The antiarrhythmic effect of ischaemic preconditioning in isolated rat hearts involves a pertussis toxin sensitive mechanism. *Cardiovasc. Res.* **27**, 674–680.

ventricular fibrillation occurred during ischaemia or during reperfusion. Further, the ultimate infarct size in the control group of dogs in this study was extremely variable, ranging from about 3% to around 34% of the risk region. These results have not as yet been published in full but it is difficult to accept the conclusion that "preconditioning failed to attenuate the incidence of lethal ventricular fibrillation in pentobarbitone-anaesthetized dogs" (Przyklenek and Kloner 1995), especially as the group size was small. Another difference of some importance is the fact that, because the main aim of these studies was to examine infarct size, multiple preconditioning occlusions were used, usually four 5 min episodes. This might well have detrimental effects on arrhythmias during a subsequent prolonged occlusion. Indeed, in the study from Lucchesi's group (Li *et al.*, 1990), whereas a single preconditioning occlusion (of 5 min) reduced the incidence of ventricular fibrillation, a greater number of preconditioning occlusions (12 x 5 min) resulted in a high mortality during the subsequent prolonged occlusion. There is indeed recent direct evidence that repeated, brief coronary artery occlusions lead to progressive deterioration of cardiac myocyte function especially at the level of the mitochondria (Pomar *et al.*, 1995). It would be of interest to determine in a model which consistently demonstrates an anti-arrhythmic effect of preconditioning (e.g. Vegh *et al.*, 1990, 1992a,b) whether multiple occlusions still protect the myocardium against ischaemia-induced life-threatening ventricular arrhythmias in a similar way to a single, or two such occlusions. It would certainly be difficult in this model to reproduce a protocol in which the duration of the preconditioning occlusions was as long as 10 min; reperfusion after such long preconditioning occlusions would, in our hands, certainly result in immediate ventricular fibrillation.

Of course, there are difficulties involved in the use of dogs to assess ventricular arrhythmias during coronary artery occlusion. The severity of such arrhythmias depends upon heart size, the site of occlusion, the area at risk, the extent of the pre-existing coronary collateral circulation and the degree of sympathetic nervous system activity. These factors are not always easy to control. Perhaps the most important of these factors are the extent of the pre-existing coronary collateral circulation and the choice of anaesthetic. The importance of pre-existing coronary collaterals is well illustrated by the studies of Meesmann (1982). This is illustrated in *Table 1* and in *Figure 3*. One explanation therefore for the anti-arrhythmic effects of ischaemic preconditioning in dogs would be the opening up of pre-existing collateral vessels by the short preconditioning

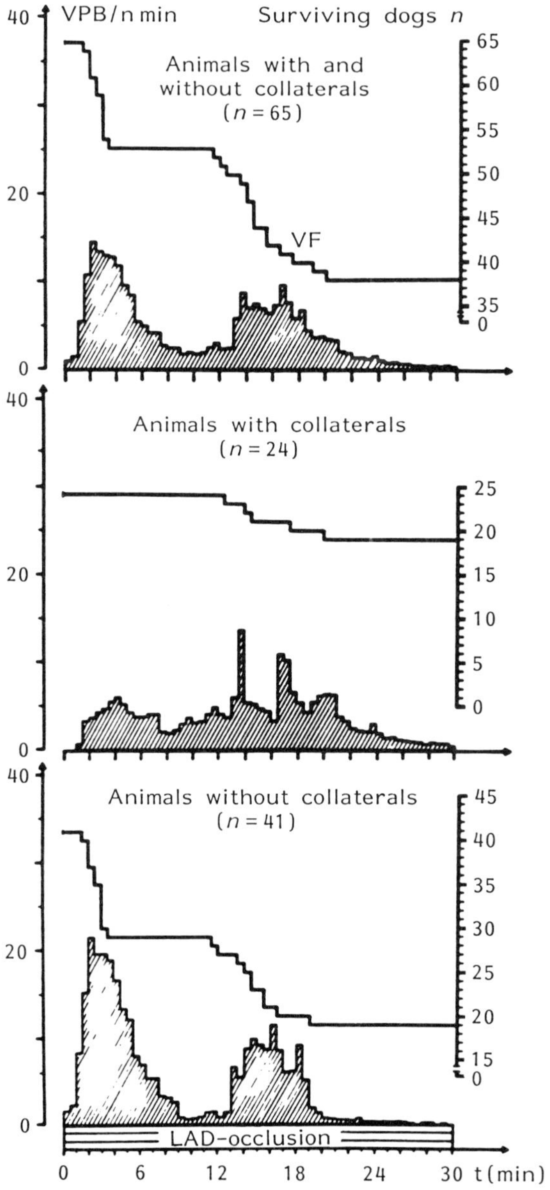

Figure 3. The incidence of ventricular premature beats (VPB) and of ventricular fibrillation (VF) during a 30 min occlusion of the left anterior descending coronary artery (LAD) in a group of 65 mongrel dogs (top figure). When these animals are subdivided into those with, and those without, pre-existing coronary collaterals, a marked difference is observed with regard to the incidence of VF and the distribution of VPBs. There is a clear differentiation to two distinct periods of VPBS (phase IA and phase IB) especially in those dogs without pre-existing collaterals. Reproduced from Meesmann (1982) in *Early Arrhythmias Resulting from Myocardial Ischaemia*, Macmillan Press Ltd, Basingstoke.

Table 1. Incidence of ventricular fibrillation after acute coronary occlusion; dependence on coronary collaterals

	Total	Ventricular fibrillation	
	n	*n*	per cent
LAD artery			
With collaterals	7	0	0
Without collaterals	9	5	55.6
LFCX artery			
With collaterals	27	1	3.7
Without collaterals	19	19	100

$p < 0.001$.
Reproduced from Meesmann (1982) in *Early Arrhythmias Resulting from Myocardial Ischaemia*, Macmillan Press Ltd, Basingstoke.

occlusions. However, there is no evidence that coronary blood flow is higher when the artery is occluded a second time, that an increase in collateral or overlap blood flow is responsible for the decreased injury seen at sites immediately below surface electrodes (Barber 1983; Li *et al.*, 1990) or that there is any significant change in peripheral coronary pressure during successive coronary occlusions (Vegh and Parratt, unpublished obervations). Further, the phenomenon of preconditioning is present in those species with very little coronary collateral development (rats, rabbits) in which a permanent occlusion leads quickly to substantial, whole wall thickness ischaemic injury.

The choice of anaesthetic is certainly important (Marshall and Parratt, 1980). This may be due in part to ensuing variations in autonomic tone. Acute myocardial ischaemia elicits both sympathetic and parasympathetic (vagal) reflexes and a dominant sympathetic activation would lead to increases in heart rate and myocardial oxygen consumption and to a decrease in cardiac electrical stability, facilitating the occurrence of malignant arrhythmias. It is well established clinically that acute psychological events can act as predictors of cardiac arrhythmias and that these effects may be related to sympathetic nervous system discharge. There is good evidence too that lateral asymmetry in the sympathetic control of the heart also leads to increased ventricular arrhythmias which can be suppressed by β-adrenoceptor blockade. If, as has been suggested (Parratt and Piacentini, 1993), the anti-arrhythmic effects of ischaemic preconditioning are due to modulation of cardiac sympathetic activity, then it would not be surprising if this phenomenon was more apparent under conditions of high sympathetic drive, such as would occur under chloralose–urethane anaesthesia. One would presume that this would be less marked in dogs anaesthetized with barbiturates.

It would be surprising if the suppression of arrhythmias, so well demonstrated in various rat models, was not part of the general protective mechanisms associated with ischaemic preconditioning. The most likely explanation for this anti-arrhythmic effect, as with the reduction in infarct size and the enhanced recovery of contractile function following reperfusion, is a reduction in the severity of ischaemia. This is clear from epicardial ST-segment mapping studies and from the measurement of changes of inhomogeneity of activation within the ischaemic area (e.g. Vegh *et al.*, 1990, 1992a).

4. The anti-arrhythmic effects of cardiac pacing

The question as to whether it is possible to induce preconditioning of the myocardium by means other than short periods of complete coronary

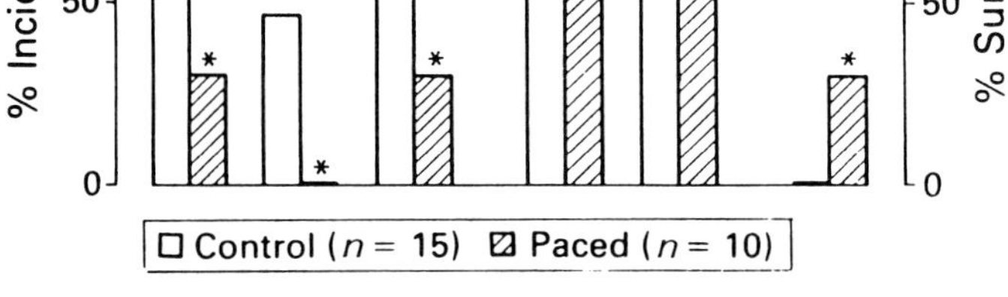

Figure 4. The severity of arrhythmias during a 25 min occlusion of the left anterior descending branch of the coronary artery in anaesthetized dogs (open columns) is markedly reduced if the occlusion is preceded by two 2 min periods of cardiac pacing (300 beats min^{-1}). There was a 5 min reperfusion time between the end of the last pacing period and the commencement of the occlusion. Pacing also increases the survival from the combined ischaemia–reperfusion insult. *$p <$ 0.05. Reproduced with permission from Vegh *et al.* (1991a) Transient ischaemia induced by rapid cardiac pacing results in myocardial preconditioning. *Cardiovasc. Res.* **25,** 1051–1053.

artery occlusion was initially investigated by rapid cardiac pacing. In anaesthetized dogs, pacing the right ventricle to a rate of 300 beats per min for two 2 min periods (with a 5 min rest period between) led to a reduced severity of arrhythmias when the coronary artery was occluded, for the first time, for a total period of 25 min (Vegh *et al.*, 1991a). The results are illustrated in *Figure 4*. A similar protection is seen if the heart is paced at a lower frequency (220 beats min^{-1}) for four 5 min periods (Kaszala *et al.*, 1995, 1996). As with the anti-arrhythmic effects of classical preconditioning, the protection depends critically upon the interval between the end of the preconditioning pacing stimulus and the commencement of the prolonged coronary artery occlusion. This is well illustrated in *Figure 5*. It is only when the coronary artery is occluded very soon after cessation of pacing that protection occurs. Thus, when the artery is occluded within 5 min of the end of the pacing stimulus, there is no ventricular fibrillation. However, this protection is lost if there is as

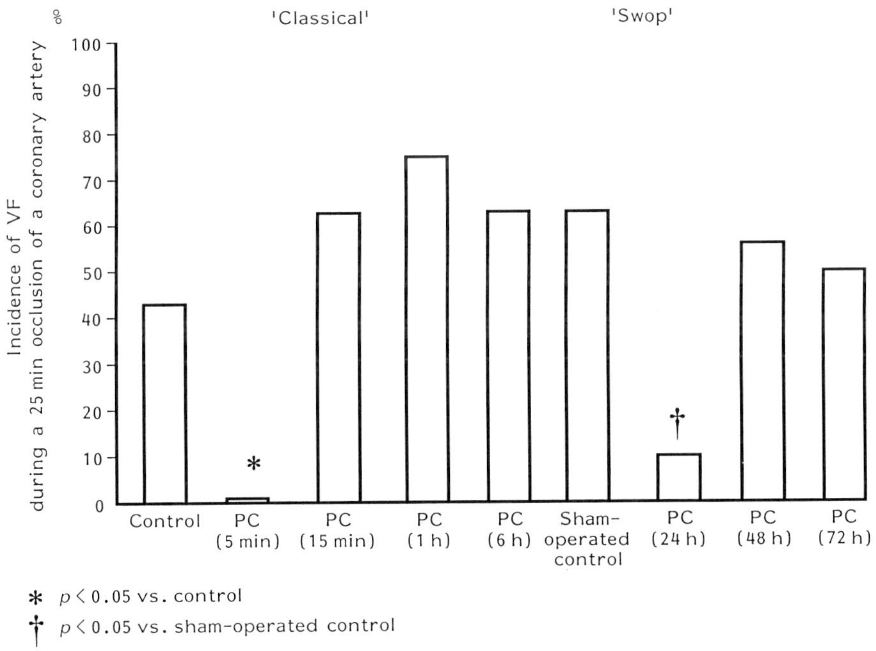

Figure 5. The incidence of ventricular fibrillation (VF) in anaesthetized, mongrel, open-chest dogs subjected to a 25 min occlusion of the left anterior descending coronary artery when that occlusion was preceded by four 5 min periods of cardiac pacing (220 beats min^{-1}). The incidence of VF is significantly reduced when the artery is occluded either 5 min or 24 h after the preconditioning pacing stimulus (PC).

little as a 15 min rest period between the end of the preconditioning stimulus and the commencement of the occlusion. Again, as with classical preconditioning, the protection is powerful (absence of ventricular fibrillation), but transient. Of greater interest, however, is the fact that this protection returns several hours later. This is discussed in more detail later.

We do not understand why rapid cardiac pacing protects the heart against the effects of a subsequent coronary artery occlusion. The most likely explanation is that, for some reason, the severity of ischaemia during a coronary artery occlusion is reduced by prior cardiac pacing. This is clear from measurements of ST-segment changes and of the degree of inhomogeneity of electrical activation within the ischaemic area (Kaszala *et al.*, 1996; Vegh *et al.*, 1991a). Certainly, periods of rapid cardiac pacing in a rabbit model reduce ischaemia-induced electrocardiographic changes and alterations in left ventricular filling pressure (Szekeres *et al.*, 1993). Further, measurements of ST-segment changes in the subendocardium of the left ventricle during pacing from the right ventricle demonstrate that pacing does induce transient ischaemia (Kaszala *et al.*, 1996). The more transient nature of the protection afforded by cardiac pacing, as compared to that afforded by complete coronary artery occlusions, might be explained by the less pronounced ischaemia during the preconditioning stimulus. Certainly, the optimum conditions for pacing induced cardioprotection have not as yet been elucidated.

4.1. Delayed anti-arrhythmic effects of cardiac pacing

One of the disappointments of classical ischaemic preconditioning is the transient nature of the protection. This depends on the nature of the experimental model but in most experiments the protection is lost as the time interval between the preconditioning stimulus and the prolonged ischaemic insult is increased to 1 or 2 h; the protection induced by cardiac pacing disappears in an even shorter time. It was a very important step therefore when it was discovered that the protection afforded by short periods of coronary artery occlusion returned several hours later (Kuzuya *et al.*, 1993; Marber *et al.*, 1993), a phenomenon which has been well described as the 'second window of protection' (Yellon and Baxter, 1995). They showed that brief periods of coronary artery occlusion result in both early (and transient) protection against myocardial necrosis, and delayed protection with a time course of many hours. A similar delayed

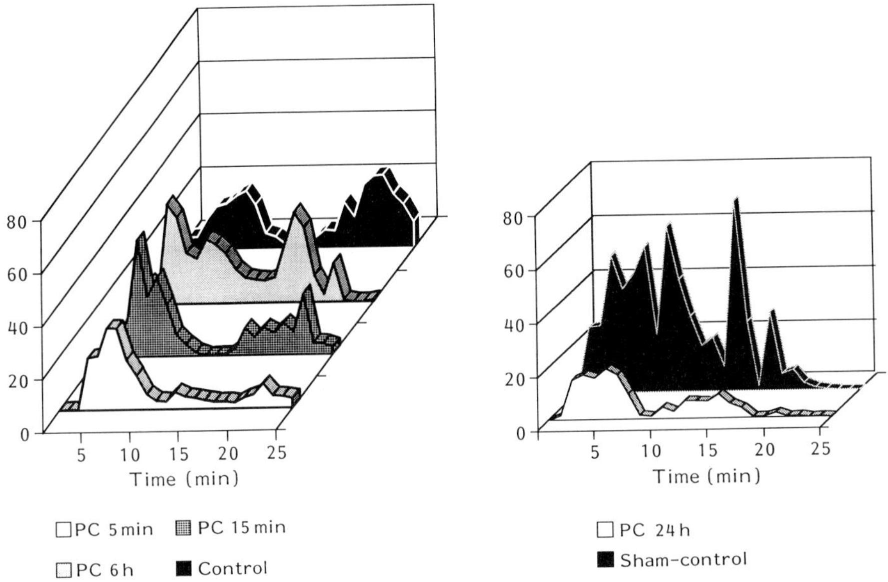

Figure 6. Changes in the distribution of ventricular premature beats during a 25 min occlusion of the left anterior descending coronary artery in anaesthetized mongrel dogs when that occlusion is preceded by four 5 min periods of cardiac pacing (220 beats min^{-1}). When the artery is occluded either 5 min or 24 h (but not 15 min or 5 h) after the end of the pacing period there is a marked suppression of ventricular ectopic activity, particularly during phase Ib and, as illustrated in *Figure 5*, a reduction in the incidence of ventricular fibrillation.

protection against ischaemia-induced arrhythmias results from cardiac pacing, and this is illustrated in *Figures 5* and *6*. The reduction in the incidence of ventricular fibrillation, which results when a coronary artery is occluded 5 min after the cessation of pacing, returns 24 h later. There is evidence (Kaszala *et al.*, 1996) that protection against ventricular arrhythmias other than fibrillation is even more prolonged, with some evidence of reduced arrhythmia severity even 72 h after the end of the pacing stimulus. Again, we know little about the optimum conditions for protection against arrhythmias by cardiac pacing, but we do now know that if the pacing stimulus is repeated during the 'fade', then the protection is even more prolonged (i.e. that repeated pacing widens the time window of the anti-arrhythmic effects; Kis *et al.*, 1996).

It is interesting to note that the phenomenon of delayed myocardial protection is not a new one. Probably the first example came from studies involving the administration of catecholamines. These studies, which were commenced in the 1950s by a group of mainly Hungarian and Czech scientists (Selye, Rona, Balazs, Poupa), were concerned with the protection afforded against toxic doses of isoprenaline by the prior administration of smaller doses of this β-adrenoceptor agonist. That this was not a simple manifestation of tolerance was shown by the fact that coronary artery occlusion also protected against the toxic effects of large doses of isoprenaline. Such myocardial protection lasted for several days or even weeks (reviewed by Parratt and Szekeres, 1995). The reverse phenomenon, protection by catecholamines against cardiogenic shock and the severity of ventricular arrhythmias resulting from coronary artery occlusion, was shown somewhat later by Beckmann and his colleagues (Beckmann *et al.*, 1981). The incidence of ventricular fibrillation during coronary embolization with microspheres was much less in adrenaline-tolerant dogs than in the controls; only one of 14 tolerant dogs fibrillated on coronary artery occlusion compared to 11 out of 32 in the control group. Since all these deaths occurred early, within the first 50 min of embolization, this suggests that the dogs were in some way protected against early (phase 1) arrhythmias. More recently, the administration of noradrenaline, phenylephrine and isoprenaline has been shown to protect against arrhythmias resulting from coronary artery occlusion 24 h later. This was associated with marked increases in *c-phos* and *c-jun* mRNA levels and increases in heat stress protein (hsp) 70 gene expression.

Similar protection against arrhythmias was also seen following the administration of prostacyclin (and its stable analogues; also reviewed by Parratt and Szekeres, 1995) and of bacterial endotoxin, or of non-toxic derivatives of lipid-A, the active part of the endotoxin molecule, such as monophosphoryl lipid-A (Vegh *et al.*, 1996). The protective effect of bacterial endotoxin against coronary artery occlusion-induced arrhythmias is illustrated in *Figure 7*. In these studies, rats were given bacterial endotoxin and, at various times after, the animals were sacrificed, the hearts removed, perfused according to Langendorff and the coronary artery occluded. This normally results in marked ventricular ectopic activity. However, in those hearts removed from rats given bacterial endotoxin the severity of these arrhythmias was much reduced, with a time course similar to that for the induction of nitric oxide synthase (Wu Song *et al.*, 1994). Similar protection is seen *in vivo* where a reduction in myocardial infarct size, as well as of arrhythmia severity, can also be demonstrated (Wu Song *et al.*, 1996).

We do not know whether the delayed protection associated with catecholamine or prostacyclin administration, endotoxaemia, cardiac

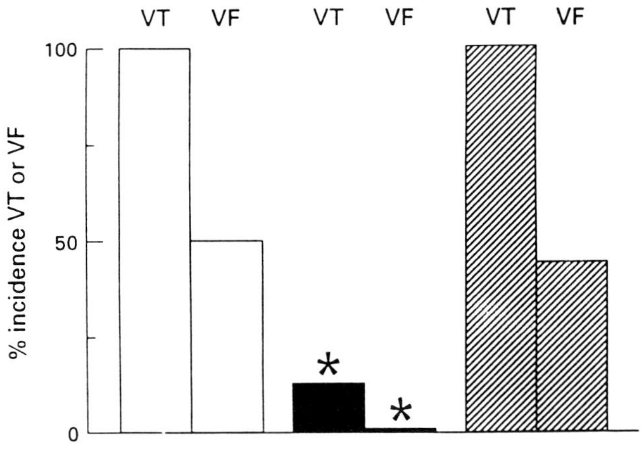

Figure 7. The incidence of ventricular tachycardia (VT) and of ventricular fibrillation (VF) in hearts removed from rats administered either saline (open columns), endotoxin (solid columns) or a combination of endotoxin and dexamethasone (hatched columns) 8 h previously and then subjected to coronary artery occlusion. There is marked reduction in the incidence of these arrhythmias in hearts removed from rats given endotoxin. This protection is abolished by the prior administration of dexamethasone. $^{*}p < 0.05$ compared to saline. Reproduced from Wu Song *et al.* (1994) with permission from the *British Journal of Pharmacology*.

pacing, heat stress, chronic hypoxia or ischaemic preconditioning have mechanisms in common. Nor whether, if not, they are additive. This has important repercussions for more prolonged myocardial protection.

5. Possible mechanisms for the anti-arrhythmic effect of ischaemic preconditioning and of cardiac pacing

The possible mechanisms involved in ischaemic preconditioning have been well reviewed elsewhere, including in this volume. Most would probably agree that the protection is triggered by the release of protective mediators from either ischaemic cardiac myocytes or endothelial cells. There are a number of such possible endogenous myocardial protective substances (Parratt, 1993). These include adenosine (reviewed by Downey in Chapter 2), bradykinin, prostanoids such as prostacyclin, and nitric oxide. It is more than likely that several of these mediators are released in response to the transient ischaemic injury which would result from preconditioning. If preconditioning really is a mechanism through which the heart is able to protect itself against ischaemic stress then it would be reasonable that more than one protective mediator is released. Such mediators, acting on different receptors, might induce protection in different ways or there might be a final, common pathway. Some believe that this involves the translocation of protein kinase C (PKC) from the cytosol to the sarcolemmal and nuclear membranes. Since this possibility is reviewed in detail elsewhere we will concentrate on another, alternative hypothesis for which there is rather less evidence.

This alternative hypothesis is illustrated in *Figure 8*. It assumes that the target organ for the preconditioning stimulus is the endothelium and

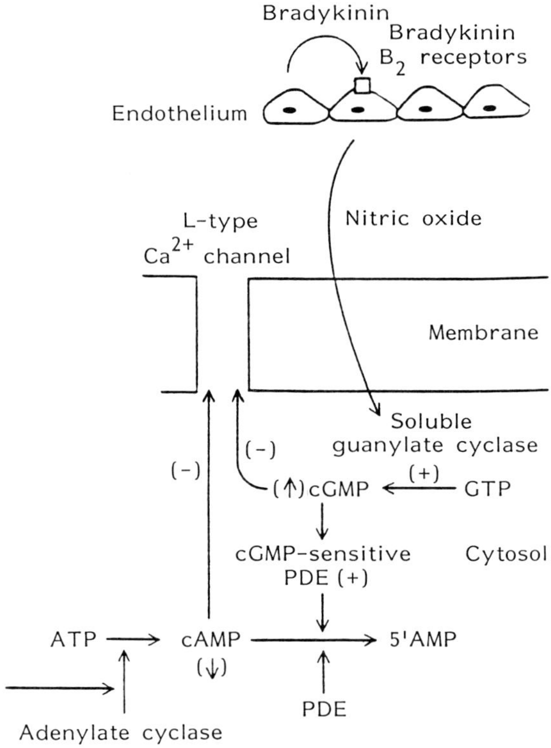

Figure 8. The alternative hypothesis for the anti-arrhythmic effects of ischaemic preconditioning. Bradykinin is released, probably from endothelial cells (which have the capacity for both generating and releasing kinins) which then acts on B_2 receptors on the endothelial surface to increase the calcium transient within these cells and activate the L-arginine nitric oxide pathway. Nitric oxide then communicates with cardiac myocytes, stimulates soluble guanylyl cyclase and elevates cyclic GMP. This stimulates a cGMP-sensitive phosphodiesterase, thus reducing cAMP levels, inhibits calcium entry through L-type calcium channels and depresses myocardial contractility. For discussion see text.

that mediators, derived from endothelial cells, modify arrhythmogenesis by direct communication, via diffusable mediators, with cardiac myocytes. The current hypothesis, for which there is some evidence, suggests that the early release of bradykinin is important, that this triggers the release from endothelial cells of both prostacyclin and nitric oxide and that it is these mediators that lead to myocyte protection. With nitric oxide the most likely scenario is that there is an elevation of cyclic guanosine monophosphate (cGMP) within the myocardium. This could modify arrhythmogenesis by suppressing calcium entry into the cell through L-type channels, by reducing levels of cyclic adenosine monophosphate (cAMP; by stimulating a cGMP-sensitive cAMP phosphodiesterase) and by direct depression of cardiac myocytes resulting in reduced oxygen demand. We know, for example, that nitric oxide is one of the factors involved in myocardial depression associated with sepsis and with elevated circulating levels of tumour necrosis factor (reviewed by Parratt, 1995a). Indeed, one recent suggestion for part of the myocardial depression associated with congestive heart failure is the release of tumour necrosis factor and the induction of nitric oxide synthase within cardiac myocytes (McCarthy, 1996).

The evidence for the above hypothesis can be summarized as described in Sections 5.1–5.6.

5.1. The relation between endothelial cells, endothelium-derived mediators and arrhythmogenesis

Although there is a good deal of clinical evidence that endothelial dysfunction occurs in patients with hypertension, atherosclerosis, ventricular hypertrophy and ischaemic heart disease (reviewed by Selwyn *et al.*, 1994), the relationship between endothelial dysfunction and arrhythmia severity in such patients is more difficult to document. This relationship is also difficult to demonstrate in most experimental animal models. For example, although most of the evidence suggesting endothelial involvement in the anti-arrhythmic effects of preconditioning, which we summarize below, has been obtained in a canine model, it is not possible to determine in this model how significant coronary vascular endothelial dysfunction would modify arrhythmias occurring during myocardial ischaemia. However, this has been attempted recently in rat isolated perfused hearts (Fatehi-Hassanabad *et al.*, 1996). Endothelial dysfunction was induced using Triton X100; this reversed the normal endothelium-dependent coronary vasodilator effect of bradykinin to coronary vasoconstriction. When the coronary artery was occluded in such isolated hearts with endothelial dysfunction the ventricular arrhythmias were much more severe than in control hearts with intact endothelial function. Thus, the total number of ventricular premature beats in control hearts during a 30 min period of ischaemia was 255 ± 39; this was increased to 2724 ± 34 in hearts with endothelial dysfunction. Further, the incidence and duration of ventricular tachycardia was markedly increased (e.g. from 14 ± 5.4 sec to 711 ± 186 sec). These hearts were also more sensitive to reperfusion, with 40% of the hearts with coronary vascular endothelial dysfunction showing tachyarrhythmias on reperfusion; this contrasted with the absence of such arrhythmias in the control hearts.

There is also evidence that some substances derived from the endothelium reduce arrhythmia severity. Bradykinin was infused locally into a side branch of the left anterior descending coronary artery which was then occluded. The severity of arrhythmias during the occlusion period in these bradykinin-treated dogs was very much less than in the controls, with a complete absence of ventricular tachycardia and ventricular fibrillation (Vegh *et al.*, 1991b). This is illustrated for a canine arrhythmia model in *Figure 9*. This protective effect of bradykinin is nitric oxide mediated since, when the formation of nitric oxide from its precursor L-arginine is prevented, the protective effects of bradykinin are largely abolished (Vegh *et al.*, 1993).

5.2. Endothelium-derived mediators and ischaemic preconditioning

The evidence for the hypothesis, outlined above, that substances derived from endothelial cells are involved in the anti-arrhythmic effects of ischaemic preconditioning, is circumstantial. It depends on the use of drugs to interfere at different sites between the generation of bradykinin from endothelial cells and the increase in cGMP within cardiac myocytes. The evidence is thus indirect; studies to directly determine the release of nitric oxide and of bradykinin under conditions of preconditioning are in progress.

5.3. Evidence for a role for bradykinin in ischaemic preconditioning

The activation of bradykinin B_2 receptors is responsible for the release of prostacyclin and the generation of nitric oxide by bradykinin (Hecker *et al.*, 1992). When a selective antagonist of bradykinin at these receptors

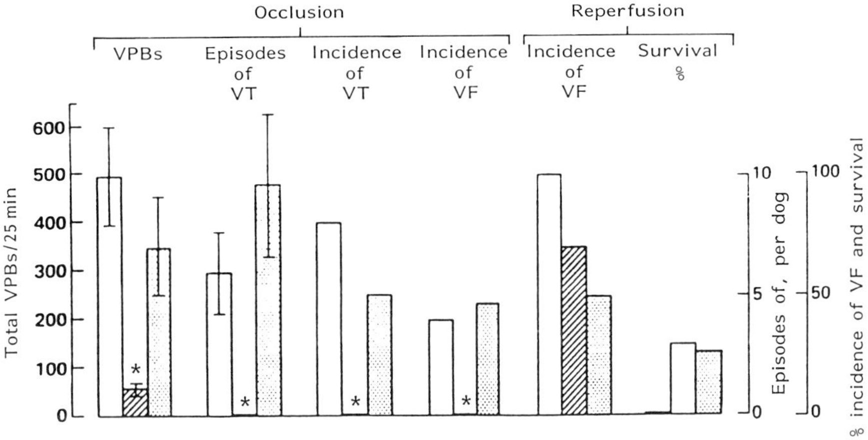

Figure 9. Suppression by the local intracoronary infusion of bradykinin [25 ng kg^{-1} min^{-1}, lined (second) columns] of the ventricular arrhythmias that occur in dogs subjected to a 25 min occlusion of the left anterior descending coronary artery. The control data are shown in the open columns. All aspects of arrhythmia severity are reduced or abolished by bradykinin [e.g. ventricular premature beats (VPBs, number of episodes and incidence of ventricular tachycardia (VT) and incidence of ventricular fibrillation (VF)]. This anti-arrhythmic effect of locally infused bradykinin is not seen if the L-arginine nitric oxide pathway is blocked by N^G-nitro-L-arginine methyl ester [L-NAME; 5 mg kg^{-1}, stippled (third) columns]. The beneficial effect of bradykinin on survival from the combined ischaemia–reperfusion insult is not influenced by inhibition of nitric oxide production. Reproduced from Vegh *et al.* (1993) with permission from the *British Journal of Pharmacology.*

(HOE 140, icatibant) is given prior to the preconditioning stimulus, then there is a greatly increased incidence of arrhythmias during the brief preconditioning coronary artery occlusions. Thus, there were more ventricular premature beats (44 ± 12 vs. 10 ± 3), a high incidence of ventricular fibrillation (50% vs. 21%) and a more pronounced ST-segment elevation (recorded from epicardial electrodes) during the preconditioning procedure in dogs given icatibant (Vegh *et al.*, 1994a). Icatibant also reversed most of the protective (anti-arrhythmic) effects of preconditioning (reduced number of ventricular premature beats, decreased incidence, and number of episodes, of ventricular tachycardia). However, the incidence of ventricular fibrillation was unaffected. Further, survival from the combined ischaemia–reperfusion insult was much less in those dogs preconditioned in the presence of icatibant (Vegh *et al.*, 1994a). The conclusion from these experiments was that bradykinin acted as a 'trigger' for preconditioning. Since icatibant also reduced the protection afforded by preconditioning when given prior to the prolonged coronary artery occlusion, it appears to function also as one of the protective anti-arrhythmic 'mediators' generated by the heart under these conditions. Because, as indicated above, in this model the anti-arrhythmic effects of bradykinin itself are mediated by nitric oxide, much of the protection afforded by preconditioning against ventricular arrhythmias must also result from nitric oxide generation.

5.4. Evidence for a role for nitric oxide in ischaemic preconditioning

The approach was to determine whether inhibition of constitutive nitric oxide synthase (with various analogues of L-arginine), or inhibition of guanylyl cyclase, the target for nitric oxide in cardiac myocytes, modified the protective anti-arrhythmic effects of ischaemic preconditioning. Again, either prevention of nitric oxide generation, or inhibition of its effects on guanylyl cyclase, attenuated the protective effects of ischaemic preconditioning (Vegh *et al.*, 1992b,c). This is illustrated for methylene blue, an inhibitor of guanylyl cyclase, in *Figure 10*. The local intracoronary administration of methylene blue, given during both the preconditioning procedure and the prolonged occlusion, abolished the protective effect of preconditioning on arrhythmia severity.

5.5. Measurement of cGMP within the ischaemic myocardium in preconditioned dogs

There is preliminary evidence (Parratt, 1995b) that cGMP levels are significantly higher in the ischaemic region of preconditioned dogs than in adjacent non-ischaemic areas of the left ventricular wall. Again, these results are not conclusive; what we need is a study of the time course of cAMP and cGMP levels in ischaemic and non-ischaemic regions of the left ventricular wall during coronary artery occlusion in dogs subjected to ischaemic preconditioning.

5.6. Cyclooxygenase products with particular reference to prostacyclin

Prostacyclin is generated under conditions of ischaemia and the amount released is related to the severity of early (phase 1a) arrhythmias (Coker

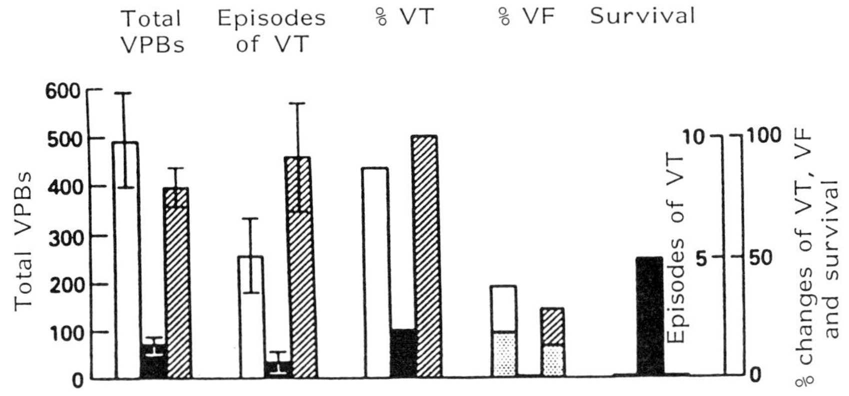

Figure 10. The effect of infusing methylene blue (by intracoronary infusion and given both during preconditioning and the prolonged occlusion in a total dose of 325 mg; shaded columns) on the protective effects of ischaemic preconditioning (solid columns) in anaesthetized mongrel dogs. The control data are seen in the initial open columns. Shown are the total number of ventricular premature beats during the 25 min occlusion period (VPBs) the number of episodes and incidence of ventricular tachycardia (VT), the incidence of ventricular fibrillation (VF) and survival from the combined ischaemia–reperfusion insult. The incidence of VF is given both as the total incidence throughout the 25 min occlusion period and during the first 5 min (stippled column). Reproduced from Vegh *et al.* (1992c) with permission from the *British Journal of Pharmacology*.

et al., 1981). Further, potentiation of the effects of released prostacyclin (e.g. by inhibiting its breakdown with nafazatrom) is profoundly anti-arrhythmic (Coker and Parratt, 1984). The evidence that cyclooxygenase products such as prostacyclin, presumably released from the endothelium, are involved in the anti-arrhythmic effects of preconditioning is again, indirect. If the cyclooxygenase pathway is blocked, for example with sodium meclofenamate, then the protective effects of ischaemic preconditioning are markedly attenuated (Vegh *et al.*, 1990). Unfortunately, this approach prevents the formation of all arachidonic acid derivatives. Thus, not only is the synthesis of potentially protective substances such as prostacyclin inhibited, so also is the formation of substances, like thromboxane, which have detrimental effects under conditions of myocardial ischaemia (reviewed by Parratt, 1989). It is not possible to inhibit prostacyclin formation and release selectively.

The above scenario suggests that a number of protective mediators are released from endothelial cells which are able to profoundly modify arrhythmia severity. These substances do not include adenosine since blockade of appropriate adenosine receptors does not influence the anti-arrhythmic effects of ischaemic preconditioning (Vegh *et al.*, 1995a). Evidence for the 'multimediator' approach is demonstrated by experiments which attempt to block simultaneously *both* the cyclooxygenase and the L-arginine nitric oxide pathways. Such dual blockade completely prevents the protective effects of ischaemic preconditioning in the canine model (Kis *et al.*, 1995).

Precisely how these mediators modify arrhythmogenesis is not known. The possibilities, as indicated above, are inhibition of calcium entry, reduced myocardial energy demand and a decrease in cAMP levels. There is another possibility which remains to be explored. Most of the mediators outlined above have in common the ability to inhibit the release of noradrenaline from cardiac sympathetic nerves by actions on pre-synaptic receptors. Because of the abundant evidence (recently reviewed by Zehender *et al.*, 1994) for an important role for released noradrenaline in the genesis of early (especially phase Ia) ventricular arrhythmias following coronary artery occlusion, it remains a possibility that the anti-arrhythmic effects of ischaemic preconditioning are mediated through inhibition, by endothelial-derived substances, of noradrenaline release.

6. Delayed myocardial protection by rapid cardiac pacing – possible mechanisms

Prompted by the studies of the Japanese and Hatter Institute groups, and by our own work on bacterial lipopolysaccharide (endotoxin), we have examined whether transient periods of ischaemia, induced by rapid cardiac pacing, protect the myocardium against arrhythmias resulting when a coronary artery is occluded several hours later. The *early* protection afforded by cardiac pacing is transient. It is even more short-lived than that induced by brief periods of coronary artery occlusion, since it is lost within 15 min of the cessation of pacing (*Figure 6*). The return of protection around 20–24 h after the pacing stimulus bears striking similarities to the time course outlined for the second window of protection described by Yellon and his colleagues (Marber *et al.*, 1993, and reviewed by Yellon and Baxter, 1995). The protection against ventricular fibrillation is particularly marked at this time; even 48 and 72 h after the cessation of pacing there are fewer ventricular premature beats and episodes of ventricular tachycardia when the anterior descending coronary artery is occluded. Of particular interest, and perhaps importance, is the fact that re-introducing the pacing stimulus prolongs the protection significantly (Kis *et al.*, 1996).

The mechanisms of this protection are obscure but the results of two studies suggest that mediators similar to those involved in classical ischaemic preconditioning play an important role. Thus pacing in the presence of icatibant fails to protect the myocardium when the coronary artery is occluded 20 h later (Vegh *et al.*, 1995b) and the protection is also completely lost if dexamethasone is administered before the pacing stimulus (Vegh *et al.*, 1994b). This might suggest a role for nitric oxide generated by nitric oxide synthase induced perhaps in cardiac myocytes. Thus the pronounced anti-arrhythmic and anti-ischaemic effects of bacterial endotoxin, almost certainly due to the induction of nitric oxide synthase and cyclooxygenase 2, are likewise not seen in animals pretreated with dexamethasone (Wu Song *et al.*, 1994) or by cycloheximide, an inhibitor of protein synthesis. Although there is no evidence that this degree of cardiac pacing in dogs induces nitric oxide synthase in the heart there is evidence that chronic exercise in this species does increase coronary vascular nitric oxide production and nitric oxide synthase gene expression (Sessa *et al.*, 1993) and also augments the sensitivity of canine coronary arteries to endothelium-dependent effects of bradykinin (Mombouli *et al.*, 1996).

7. Conclusions

It seems clear that one of the important, perhaps the most important, manifestations of the protection afforded by ischaemic preconditioning is suppression of life-threatening ventricular arrhythmias occurring during myocardial ischaemia, and perhaps also during reperfusion. This has been demonstrated in at least three species, both *in vivo* and *in vitro*, and seems to be an inherent property of the heart itself. There is some evidence that this protection resides, at least in part, in endothelial cells since endothelial dysfunction increases the severity of ventricular arrhythmias occurring during myocardial ischaemia. There is some evidence that substances derived from vascular endothelium such as bradykinin, nitric oxide and prostacyclin play a role in this protection. However, at present, the evidence is indirect and depends upon the use of drugs that block various sites in the pathway leading to the elevation of cGMP within cardiac myocytes. Some of these drugs (e.g. methylene blue) are not sufficiently specific. Of particular interest is the fact that there is a 'second window of protection' for the anti-arrhythmic effects of ischaemic preconditioning, as there is for the reduction in myocardial ischaemic damage, and that the time course of this protection (maximal protection around 24 h but with some protection still present 96 h after the preconditioning stimulus) is similar to that described for the 'second window'. The mechanisms of this delayed protection are poorly defined but may also involve the generation of endogenous myocardial protective substances from the endothelium and/or increased endothelial receptor sensitivity.

Acknowledgements

The work described in this chapter has been variously supported, since 1989, by the Royal Society, the Wellcome Trust, the Scottish Home and Health Department, the British Council (in collaboration with the Hungarian Committee for Technical Development) and by the Hungarian State Government (OTKA). The collaboration has also been stimulated by a Scientific Network Grant from the European Union (No. CIPA-CT-92-4009) and by a European Union Biomed I grant (No. BMHI-CT92-1893). We particularly acknowledge the collaboration of many individuals within the Szeged and Glasgow departments especially Professor Laszlo Szekeres, Dr Eva Udvary and Dr Nigel Pyne and, in Strasbourg, Dr Claire Lugnier and Professor Jean-Claude Stoclet. It is a pleasure also to acknowledge the significant help, in the preparation of the manuscript, of Mrs Margaret Laird.

References

Barber MJ. (1983) Effect of time interval between repeated brief coronary artery occlusions on arrhythmias, electrical activity and myocardial blood flow. *J. Am. Coll. Cardiol.* **2,** 699–705.

Beckman CB, Niazi Z, Dietzman RH and Lillehei RC. (1981) Protective effects of epinephrine tolerance in experimental cardiogenic shock. *Circul. Shock* **8,** 137–149.

Coker SJ and Parratt JR. (1984) The effects of nafazatrom on arrhythmias and prostanoid release during coronary artery occlusion and reperfusion in anaesthetised greyhounds. *J. Mol. Cell. Cardiol.* **16,** 43–52.

Coker SJ, Parratt JR, Ledingham IMcA and Zeitlin IJ (1981) Thromboxane and prostacyclin release from ischaemic myocardium in relation to arrhythmias. *Nature* **291,** 323–324.

Fatehi-Hassanabad Z, Furman BL and Parratt JR. (1996) The effect of the endothelium on coronary artery occlusion-induced arrhythmias in rat isolated perfused hearts. *J. Physiol.*, in press.

Gulker H, Kramer B, Stephan K and Meesmann W. (1977) Changes in ventricular fibrillation threshold during repeated short-term coronary occlusion and release. *Basic Res. Cardiol.* **72,** 547–562.

Harris AG. (1950) Delayed development of ventricular ectopic rhythms following experimental coronary occlusion. *Circulation* **1,** 1318–1326.

Hecker M, Dambacher T and Busse R. (1992) Role of endothelium-derived bradykinin in the control of vascular tone. *J. Cardiovasc. Pharmacol.* **20** (suppl. 9), 555–561.

Kaszala K, Vegh A, Parratt JR and Papp JGy. (1995) Time course of pacing induced preconditioning in dogs. *J. Mol. Cell. Cardiol.* **27,** A145.

Kaszala K, Vegh A, Papp J Gy and Parratt JR. (1996) Time-course of the protection against ischaemia and reperfusion-induced ventricular arrhythmias resulting from brief periods of cardiac pacing. *J. Mol. Cell. Cardiol.*, in press.

Kis, A, Vegh A, Papp J Gy and Parratt JR. (1995) Dual blockade of the cyclooxygenase and L-arginine nitric oxide pathways prevents the antiarrhythmic effect of preconditioning. *J. Mol. Cell. Cardiol.* **27,** A159.

Kis A, Vegh A, Papp JGy and Parratt JR. (1996) Repeated pacing widens the time window of delayed protection against ventricular arrhythmias in dogs. *J. Mol. Cell. Cardiol.*, in press.

Komori S, Parratt JR, Szekeres L and Parratt JR. (1990a) Preconditioning reduces the severity of ischaemia and reperfusion-induced arrhythmias in both anaesthetised rats and dogs. *J. Physiol.* **423,** 16P.

Komori S, Fujimaki S, Ijili H, Asakawa T, Watanabe Y, Tamura Y and Parratt JR. (1990b) Inhibitory effect of ischemic preconditioning on ischemic arrhythmias using a rat coronary artery ligation model. *Jpn. J. Electrocardiol.* **10,** 774–782.

Kuzuya T, Hoshida S, Yamashita N, Fuji H, Oe H, Hori M, Kamada T and Tada M. (1993) Delayed effects of sublethal-ischemia on the acquisition of tolerance to ischemia. *Circ. Res.* **72,** 1293–1299.

Lawson CS, Coltart DJ and Hearse DJ. (1993a) 'Dose-dependency' and temporal characteristics of protection by ischaemic preconditioning against ischaemia-induced arrhythmias in rat hearts. *J. Mol. Cell. Cardiol.*, **25,** 1391–1402.

Lawson CS, Avkiran M, Shattock MJ, Coltart DJ and Hearse DJ. (1993b) Preconditioning and reperfusion arrhythmias in the isolated rat heart: true protection or temporal shift in vulnerability? *Cardiovasc. Res.* **27,** 2274–2281.

Li Y and Kloner RA. (1992) Cardioprotective effects of ischemic preconditioning not mediated by prostanoids. *Cardiovasc. Res.* **26,** 226–231.

Li GC, Vasquez JA, Gallagher KP and Lucchesi BR. (1990) Myocardial protection with preconditioning. *Circulation* **82,** 609–619.

Marber MS, Latchman DS, Walker JM and Yellon DM. (1993) Cardiac stress protein elevation 24 hours after brief ischemia or heat stress is associated with resistance to myocardial infarction. *Circulation* **88,** 1264–1272.

Maroko PR, Kjekshus JK, Sobel BE, Watanabe T, Covell JW, Ross J and Braunwald E. (1971) Factors influencing infarct size following experimental coronary artery occlusions. *Circulation* **43,** 67–74.

Marshall RJ and Parratt JR. (1977) The haemodynamic and metabolic effects of MG8926, a prospective antidysrhythmic and antianginal agent. *Br. J. Pharmacol.* **59,** 311–322.

Marshall RJ and Parratt JR. (1979) The effects of disopyramide phosphate on early post-coronary artery ligation dysrhythmias, and on epicardial ST-segment elevation, in anaesthetised dogs. *Br. J. Pharmacol.* **66,** 241–250.

Marshall RJ and Parratt JR. (1980) The early consequences of myocardial ischaemia and their modification. *J. Physiol. (Paris)* **76,** 699–715.

McCarthy M. (1996) Is there more to heart failure than a bad pump? *Lancet* **347,** 110.

Meesmann W. (1982) Early arrhythmias and primary ventricular fibrillation after acute myocardial ischaemia in relation to pre-existing coronary collaterals. In: *Early Arrhythmias Resulting from Myocardial Ischaemia* (ed. JR Parratt). Macmillan, London, pp. 93–112.

Mombouli J-V, Nakashima M, Hamra M and Vanhoutte PM. (1996) Endothelium-dependent relaxation and hyperpolarisation evoked by bradykinin in canine coronary arteries: enhancement by exercise-training. *Br. J. Pharmacol.* **117,** 413–418.

Murry CE, Jennings RB and Reimer KA. (1986) Preconditioning with ischemia: a delay of lethal cell injury in ischaemic myocardium. *Circulation* **74,** 1124–1136.

Parratt JR. (1989) Eicosanoids and arrhythmogenesis. In: *Handbook of Experimental Pharmacology,* Vol. 89, *Antiarrhythmic Drugs* (ed. EM Vaughan-Williams). Springer, Berlin, pp. 569–589.

Parratt JR. (1993) Endogenous myocardial protective (antiarrhythmic) substances. *Cardiovasc. Res.* **27,** 693–702.

Parratt JR. (1995a) Nitric oxide and cardiovascular dysfunction in sepsis and endotoxaemia – an introduction and an overview. In: *Shock, Sepsis and Organ Failure, Fourth Bernard Wiggers Conference* (eds G Schlag and H Redl). Springer, Berlin, pp. 1–21.

Parratt JR. (1995b) Possibilities for the pharmacological exploitation of ischaemic preconditioning. *J. Mol. Cell. Cardiol.* **27,** 991–1000.

Parratt JR and Piacentini L. (1993) G-proteins and preconditioning. *Cardiovasc. Res.* **27,** 690–691.

Parratt JR and Szekeres L. (1995) Delayed protection of the heart against ischaemia. *Trends Pharmacol. Sci.* **16,** 351–355.

Parratt JR and Vegh A. (1994) Pronounced antiarrhythmic effects of ischaemic preconditioning. *Cardioscience* **5,** 9–18.

Piacentini L, Wainwright CL and Parratt JR. (1993) The antiarrhythmic effect of ischaemic preconditioning in isolated rat hearts involves a pertussis toxin sensitive mechanism. *Cardiovasc. Res.* **27,** 674–680.

Pomar F, Cosin J, Portoles M, Faura M, Renau-Piqueras J, Hernandiz A, Andres F, Colomer JL and Graullera B. (1995) Functional and ultrastructural alterations of canine myocardium subjected to very brief coronary occlusions. *Eur. Heart J.* **60,** 1482–1490.

Przyklenk K and Kloner RA. (1994) Preconditioning does not attenuate ventricular ectopy in the canine heart. *J. Mol. Cell. Cardiol.* **26,** 137.

Przyklenk K and Kloner RA. (1995) Preconditioning: a balanced perspective. *Br. Heart J.* **74,** 575–577.

Selwyn AP, Anderson T, Meredith I, Yeung A and Ganz P. (1994) Vascular mechanisms leading to myocardial ischemia and their relation to hypertension, left ventricular hypertrophy and arrhythmias. In: *Myocardial Ischemia and Arrhythmia* (eds M Zehender, T Mainertz and H Just). Steinkopff, Darmstadt, pp. 205–211.

Sessa WC, Pritchard K, Seyedi N, Wang J and Hintze TH. (1993) Chronic exercise in dogs increases coronary vascular nitric oxide production and endothelial cell nitric oxide synthase gene expression. *Circ. Res.* **74,** 349–353.

Shiki K and Hearse DJ (1987) Preconditioning of ischemic myocardium; reperfusion-induced arrhythmias. *Am. J. Physiol.* **253,** H1470–H1476.

Szekeres L, Papp J Gy, Szilvassy Z, Udvary E and Vegh A. (1993) Moderate stress by cardiac pacing may induce both short term and long term cardioprotection. *Cardiovasc. Res.* **27,** 593–596.

Vander Heide RS, Schwartz LM, Jennings RB and Reimer KA. (1995) Effect of catecholamine depletion on myocardial infarct size in dogs: role of catecholamines in ischemic preconditioning. *Cardiovasc. Res.* **30,** 656–662.

Vegh A, Szekeres L and Parratt JR. (1990) Protective effect of preconditioning of the ischaemic myocardium involves cyclooxygenase products. *Cardiovasc. Res.* **12,** 1020–1022.

Vegh A, Szekeres L and Parratt JR. (1991a) Transient ischaemia induced by rapid cardiac pacing results in myocardial preconditioning. *Cardiovasc. Res.* **25,** 1051–1053.

Vegh A, Szekeres L and Parratt JR. (1991b) Local coronary infusions of bradykinin profoundly reduce the severity of ischaemia-induced arrhythmias in anaesthetised dogs. *Br. J. Pharmacol.* **104,** 294–295.

Vegh A, Komori S, Szekeres L and Parratt JR. (1992a) Antiarrhythmic effects of preconditioning in an anaesthetised dogs and rats. *Cardiovasc. Res.* **26,** 487–495.

Vegh A, Szekeres L and Parratt JR. (1992b) Preconditioning of the ischaemic myocardium; involvement of the L-arginine nitric oxide pathway. *Br. J. Pharmacol.* **107,** 648–652.

Vegh A, Papp JGy, Szekeres L and Parratt JR. (1992c) The local intracoronary administration of methylene blue prevents the pronounced antiarrhythmic effect of ischaemic preconditioning. *Br. J. Pharmacol.* **107,** 910–911.

Vegh A, Papp JGy, Szekeres L and Parratt JR. (1993) Prevention by an inhibitor of the L-arginine–nitric oxide pathway of the antiarrhythmic effects of bradykinin in anaesthetised dogs. *Br. J. Pharmacol.* **110,** 18–19.

Vegh A, Papp JGy and Parratt JR. (1994a) Attenuation of the anti-arrhythmic effects of ischaemic preconditioning by blockade of bradykinin B_2 receptors. *Br. J. Pharmacol.* **113,** 1167–1172.

Vegh A, Papp JGy and Parratt JR. (1994b) Prevention by dexamethasone of the marked antiarrhythmic effects of preconditioning induced 20 h after rapid cardiac pacing. *Br. J. Pharmacol.* **113,** 1081–1082.

Vegh A, Papp JGy and Parratt JR. (1995a) Pronounced antiarrhythmic effects of preconditioning in anaesthetised dogs: is adenosine involved? *J. Mol. Cell. Cardiol.* **27,** 349–356.

Vegh A, Kaszala K, Papp JGy and Parratt JR. (1995b) Delayed myocardial protection by pacing-induced preconditioning: a possible role for bradykinin. *Br. J. Pharmacol.* **116,** 288P.

Vegh A, Papp JGy, Parratt JR and Elliott GT. (1996) Pretreatment with monophosphoryl lipid-A (MPL-C) reduces ischaemia reperfusion-induced arrhythmias in dogs. *J. Mol. Cell. Cardiol.*, in press.

Wu S, Furman BL and Parratt JR. (1994) Attenuation by dexamethasone of endotoxin protection against ischaemia-induced ventricular arrhythmias. *Br. J. Pharmacol.* **113,** 1083–1084.

Wu Song, Furman BL and Parratt JR. (1996) Delayed protection by bacterial lipopolysaccharide (endotoxin) against ischaemia-induced myocardial damage and arrhythmias. *Br. J. Pharmacol.*, in press.

Yellon DM and Baxter GF. (1995) 'A second window of protection' or delayed preconditioning phenomenon: future horizons for myocardial protection. *J. Mol. Cell. Cardiol.* **27,** 1023–1034.

Zehender M, Faber T, Koscheck U, Mainertz T and Just H (1994) Ventricular tachyarrhythmias, myocardial ischemia and sudden cardiac death in patients with hypertensive heart disease. In: *Myocardial Ischaemia and Arrhythmia* (eds M Zehender, T Mainertz and H Just). Steinkoppf, Damstardt, pp. 191–204.

6

The 'second window of protection' associated with ischaemic preconditioning

G.F. Baxter and D.M. Yellon
The Hatter Institute for Cardiovascular Studies, Department of Academic and Clinical Cardiology, University College Hospital, Grafton Way, London WC1E 6AU, UK

1. Introduction

The last 2 years have brought the realization that ischaemic preconditioning of myocardium induces two distinct phases of myocardial protection. The extremely rapid appearance of an ischaemia-tolerant state is one of the most striking features of classic preconditioning. In addition to the well-researched early phase of protection (reviewed in Chapter 1) there is now increasing evidence for a delayed phase of protection, termed the 'second window of protection', which appears many hours after classic preconditioning has disappeared. These two phases of protection occurring in response to transient ischaemic stress suggest that there are at least two different endogenous routes to cytoprotection in myocardium. Both forms of adaptation may be extremely pertinent to our appreciation of the natural

Ischaemia: preconditioning and adaptation, edited by M.S. Marber and D.M. Yellon

history of ischaemic heart disease and both may provide new templates for the development of cardioprotective strategies. Furthermore, examination of these forms of adaptive cytoprotection may be relevant to other tissues and organs. In this chapter, we will concentrate specifically on the delayed form of protection.

2. Experimental evidence for and against a second window of protection

2.1. Protection against ischaemic necrosis

The second window of protection was originally described in 1993 as a period of enhanced tolerance to ischaemic necrosis 24 h after a preconditioning stimulus of brief repetitive cycles of coronary occlusion. This discovery was made independently by two groups using rabbit (Marber *et al.*, 1993) and canine (Kuzuya *et al.*, 1993) models of coronary artery occlusion.

Initially, work in our laboratory was directed at examining the hypothesis that the 70 kDa heat shock protein (hsp70i, see Section 4) is a cytoprotective protein which, when present in sufficient quantity, is able to confer resistance to ischaemia–reperfusion injury. It had been reported by others that sublethal myocardial ischaemia could increase the expression of hsp70i mRNA and protein in myocardium (Dillmann *et al.*, 1986; Knowlton *et al.*, 1991; Mehta *et al.*, 1988) and this appeared to offer a pathophysiologically relevant method of inducing hsp70i in the heart. An experiment was conducted using a rabbit model of preconditioning with four 5 min coronary artery occlusions and subsequent recovery (Marber *et al.*, 1993). Twenty-four hours following this preconditioning stimulus, the animals were subjected to a 30 min period of coronary occlusion and infarct size was determined. The percentage of infarcted myocardium within the risk zone (% I/R) was reduced from 52% to 29%, representing a 44% reduction in infarct size. This protection was very similar to the protection that was observed in a separate group of animals following prior heat stress (a treatment well known to induce myocardial protection 24 h later) in which there was a 42% reduction in I/R.

At the same time, another group observed that repetitive brief ischaemia with reperfusion caused an increase in the endogenous antioxidant capacity of canine myocardium and that this conferred enhanced tolerance to subsequent ischaemia. Hoshida *et al.* (1993) initially showed that four 5 min coronary occlusions in the dog produced a modest but statistically significant increase in mitochondrial manganese–superoxide dismutase (manganese-SOD) activity in the subepicardial and subendocardial regions of the ischaemic zone compared with the non-ischaemic zone. This work was followed by a report of a study from the same group describing the time course of infarct limitation between 3 and 24 h after preconditioning (Kuzuya *et al.*, 1993). The preconditioning protocol was identical to that used in their earlier study examining alterations in manganese-SOD and very similar to that used by Marber *et al.* (1993), namely four 5 min coronary occlusions each separated by 5 min reperfusion. There were four phases to their infarct studies. The first stage demonstrated classic preconditioning with a 63% reduction of infarction produced by a 90 min coronary occlusion. Three hours after preconditioning there was a non-significant 17% difference between I/R in preconditioned and control groups. At 12 h following preconditioning there was a non-significant 27% reduction in I/R. Twenty-four hours following preconditioning ischaemia, I/R was 46% lower in preconditioned versus control hearts, a difference that was statistically significant.

This delayed anti-infarct effect has subsequently been confirmed in further open-chest rabbit studies which are discussed in more detail below (Baxter *et al.*, 1994, 1995a,b) and also in chronically instrumented conscious rabbits (Yang *et al.*, 1996). A third species in which a second window of protection has been described is the rat. Recently, Yamashita *et al.* (1995) have presented a preliminary report that preconditioning with two 3 min coronary occlusions, each separated by 5 min reperfusion, protected against a 30 min occlusion 24 h later. I/R was reduced from 62% in sham-operated animals to 41% in preconditioned animals.

Some studies designed to examine the second window of protection have reported negative findings. The first of these was conducted by Tanaka *et al.* (1994). Rabbits were anaesthetized with pentobarbitone, ventilated using air supplemented with oxygen and a left parasternal thoracotomy was performed. Preconditioning was effected by four 5 min occlusions of an anterolateral branch of the circumflex artery and animals were recovered for 24 or 48 h. Sustained ischaemia was then induced for 30 min followed by 180 min reperfusion and infarct size assessed. Although these workers showed that this preconditioning protocol induced early classic protection (a 72% reduction in infarct size), no protection was observed 24 or 48 h later. Reasons for this absence of protection, in contrast to our earlier study (Marber *et al.*, 1993), are not immediately clear. An increase in hsp70i immunoreactivity in preconditioned hearts was observed by Tanaka *et al.* (1994) at these late time-points. This immunoreactivity was determined by a qualitative histochemical technique. The authors discussed several differences in experimental technique between their study and that of Marber *et al.* (1993), including the anaesthetic used during the preconditioning procedure (pentobarbitone vs. opiate/diazepam). Perhaps the most far-reaching difference in surgical technique was the use of a parasternal thoracotomy by Tanaka compared to the median sternotomy employed in our studies (Baxter *et al.*, 1994, 1995a,b; Marber *et al.*, 1993). The parasternal approach may produce more post-operative trauma and discomfort than the median sternotomy. A study which we undertook in chronically instrumented conscious rabbits, in which a 33% reduction in infarct size was observed in preconditioned animals, tends to suggest that the second window of protection in this species can occur independently of any confounding effects of surgery, respiration or anaesthetic (Yang *et al.*, 1995)

A further two studies, presented in preliminary form, have examined preconditioning protocols in pig myocardium and have not seen limitation of infarction 24 h later. In the first of these studies, Schaper's group (Strasser *et al.*, 1994) reported that four 5 min cycles of ischaemia failed to elicit protection against a sustained 60 min coronary occlusion 24 h later (I/R was 71.3% in controls vs. 60.4% in preconditoned, *p* not significant). It is possible that the duration of this sustained ischaemic insult in a non-collateralized species (resulting in greater than 70% infarction of the area at risk) may have been too severe for demonstration of a protective effect. It is well appreciated that, in this type of experimental model, protective interventions may only delay the onset of necrosis and not prevent necrosis absolutely. A more recent study in the pig has been more carefully designed to examine the possibility of a second window of protection. Qiu *et al.* (1995) examined the effects of repeated brief coronary occlusions (10 × 2 min) in conscious pigs 25 min (classic preconditioning) and 24 h (second window preconditioning) before a 40 min occlusion. Classic preconditioning resulted in an 80% reduction in I/R (from 45.1% to 9.2% ; $p < 0.05$). However, in the group preconditioned 24 h earlier a 26% reduction in I/R was seen that was not statistically significant (I/R 45.1% vs. 33.3%). A further group of animals preconditioned with 25 × 2 min occlusions did not display further enhancement of ischaemic tolerance at 24 h and these workers have concluded that a second window of protection against infarction does not occur in the pig. However, as we will discuss below, it is possible that 24 h after preconditioning may not be the ideal time for examination of the delayed phase of protection against infarction and it is conceivable that in the pig maximum infarct limitation may be observed beyond 24 h.

2.2. Protection against ischaemia–reperfusion arrhythmias

There is evidence also that the delayed protection conferred by preconditioning extends to other indices of ischaemic and post-ischaemic myocardial dysfunction. Recently Vegh's group reported a delayed anti-arrhythmic effect following preconditioning (Vegh *et al.*, 1994). Four 5 min periods of ventricular rapid pacing in the dog induced a biphasic pattern of protection against coronary occlusion- and reperfusion-induced tachyarrhythmias, with a second window of protection around 20–24 h after the pacing stimulus. It is also germane to this discussion that in an investigation of the second window in a conscious rabbit model we observed abolition of coronary occlusion-induced ventricular fibrillation 24 h after preconditioning (Yang *et al.*, 1995). In control rabbits, ventricular fibrillation occurred in three out of seven animals whereas none of seven preconditioned animals fibrillated ($p < 0.1$).

2.3. Protection against myocardial stunning

Another endpoint of delayed preconditioning protection, namely protection against myocardial stunning, has been described by Bolli's group (Sun *et al.*, 1995). This novel finding is particularly interesting since it describes an index of protection in the second window that is not seen in the early window, there being no conclusive evidence that classic preconditioning confers protection against myocardial stunning in the absence of tissue necrosis. These workers used chronically instrumented conscious pigs to investigate post-ischaemic myocardial dysfunction following ten 2 min coronary artery occlusions. This protocol induced severe post-ischaemic stunning (determined by systolic thickening measurement) lasting 3–4 h. However, when the protocol was repeated in the same animals 24 h later it was found that post-ischaemic recovery of thickening fraction was significantly accelerated compared with recovery following the initial protocol. Thus it would appear that the first stunning protocol preconditioned against a subsequent stunning protocol 24 h later.

2.4. Protection against hypoxia and simulated ischaemia in isolated myocytes

Further evidence for a delayed effect of preconditioning has been obtained in isolated cardiomyocyte studies which provide interesting *in vitro* models of the second window of protection. Yamashita *et al.* (1994) established that hypoxic preconditioning of rat neonatal cardiomyocytes conferred resistance to a subsequent more prolonged hypoxic period 24 h later. Exposure of myocytes to hypoxia for 3 h resulted in creatine kinase release which was significantly reduced if the myocytes were preconditioned 24 h earlier by exposure to a brief (1 h) hypoxic period. More recently we have described a model of delayed protection in neonatal rat cardiomyocytes following preconditioning with 10 mM deoxyglucose and 20 mM lactate superfusate (Cumming *et al.*, 1996a). Metabolic 'preconditioning' with simulated ischaemia for 20 min, 24 h before a 2 h simulated ischaemia period, resulted in less cell death, as assessed by trypan blue exclusion, compared to non-preconditioned cells (56% cell death in preconditioned cells vs. 69% in non-preconditioned cells). Lactate dehydrogenase release following the 'lethal' ischaemic period was also attenuated in the cells preconditioned 24 h earlier.

3. Timecourse of the second window of protection

The slow onset of the second window of protection, originally shown in the study of Kuzuya *et al.* (1993), suggested that it was likely to be of greater duration than classic preconditioning protection. Recent work from our laboratory suggests that the delayed anti-infarct effect of preconditioning in the rabbit may extend over a period of between 1 and 3 days (Baxter *et al.*, 1995b). Interestingly, the magnitude of the infarct-limiting effect in the rabbit increases as the interval between preconditioning and acute myocardial infarction is extended beyond 24 h and is maximal 48–72 h after preconditioning. By 96 h after preconditioning, however, no protection against infarction is seen (see *Figure 1*).

In an investigation of the timecourse of the rapid pacing-induced delayed protection against arrhythmias, Vegh's laboratory have extended the interval between the pacing stimulus and coronary occlusion up to 72 h and found that protection against fibrillation (and hence mortality) was almost completely lost by 48 h. However, there was

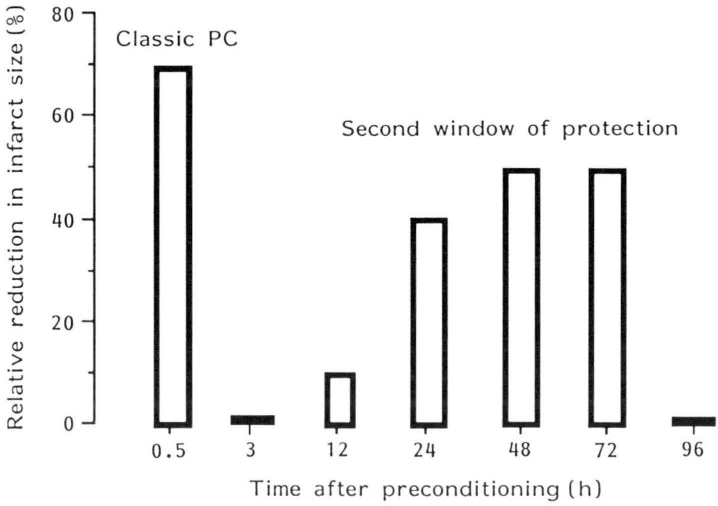

Figure 1. Composite figure derived from several experimental studies indicating the biphasic timecourse of ischaemic preconditioning against necrosis.

evidence that even 72 h later, some attenuation of parameters such as electrical inhomogeneity was present (Kaszala *et al.*, 1995).

With regard to the timecourse of the second window against myocardial stunning in the pig, Sun *et al.* (1995) initially reported that the anti-stunning effect could be renewed by preconditioning again after 24 h so that protection was seen 48 h after the first stunning protocol. No protection was observed after a 10 day interval between the preconditioning protocol and the subsequent stunning protocol. More recently, this group has examined the timecourse of this delayed anti-stunning phenomenon more specifically (Tang *et al.*, 1995). They have shown that the systolic wall thickening deficit was reduced when the second stunning protocol was induced 12 h, 24 h and 72 h after the first stunning protocol but no protection was observed 6 days after preconditioning.

4. Mechanisms of delayed protection

In considering the mechanisms through which protection is manifested in the second window an obvious possibility is that preconditioning might open coronary collateral vessels or stimulate angiogenesis, providing collateral support in the ischaemic region. However, this is unlikely to acount for the protection observed in experimental studies. In the dog, Kuzuya *et al.* (1993) observed no differences in regional myocardial blood flow (determined by radiolabelled microspheres) 24 h after four 5 min coronary occlusions, obviating the possibility that opening of preformed collaterals could account for delayed protection in this species. The rabbit is deficient in preformed collateral vessels (Maxwell *et al.*, 1987) and, moreover, it has a very limited capacity to develop collaterals in response to ischaemia. For example, Cohen *et al.* (1994) demonstrated that even after 3–4 days of repetitive coronary artery occlusion in the conscious rabbit, the ratio of collateral to normal tissue blood flow remained at a constant low level of around 5%.

The appearance of a large number of new gene products including proto-oncogenes and regulatory proteins, as well as the post-translational modification of proteins, would be anticipated after sublethal ischaemia as part of the intracellular stress response and as a consequence of the activation of various signalling pathways triggered by the local release of paracrine mediators during preconditioning. Associations between enhanced myocardial tolerance and stress-induced cytoprotective protein activity, the gradual onset of the second window of protection (Kuzuya *et al.*, 1993) and the prolonged period of protection (Baxter *et al.*, 1995a; Sun *et al.*, 1995) suggest, but do not confirm, that the phenomenon is related to changes in patterns of protein activity in the preconditioned myocardium. Almost certainly, *de novo* protein synthesis will be stimulated by ischaemic preconditioning but so far there is no evidence from studies with transcriptional and translational inhibitors *in vivo* to support directly the concept that the second window of protection involves genomic modifications. Nevertheless, certain proteins have been proposed as the putative mediators of delayed protection and changes in the content or activity of some of these proteins, notably antioxidant proteins and heat shock proteins (hsps), have been recently documented with respect to late preconditioning. We discuss below some of these potential mediators of protection and we also comment on some potential signalling pathways in this form of myocardial adaptation.

4.1. Heat shock proteins (stress proteins)

The cellular stress response which confers protection against subsequent stress is an indisputably complex phenomenon. A family of proteins known as heat shock (or stress) proteins are known to be preferentially expressed or post-translationally modified during and after metabolic and other stresses, including ischaemia. These proteins are often described according to their molecular weight [e.g. 70 kDa hsp (hsp70), 60 kDa hsp (hsp60), 27 kDa hsp (hsp27)]. Some of these proteins are expressed constitutively and others are specifically stress inducible. They have many diverse functions in the general maintenance of cellular integrity (functions often referred to as 'housekeeping' roles). These functions may include molecular chaperone activity in the control of protein synthesis, folding and degradation, and some hsps are associated with specific organelles (e.g. hsp60 and the mitochondrion, hsp27 and the cytoskeleton).

It is known that 24 h following heat stress, hearts are relatively resistant to subsequent ischaemia–reperfusion (Currie *et al.*, 1988, 1993; Donnelly *et al.*, 1992; Karmazyn *et al.*, 1990; Steare and Yellon, 1993; Walker *et al.*, 1993; Yellon *et al.*, 1992). This acquired tolerance to ischaemia–reperfusion as part of the heat stress response, is associated with increased activity and/or content of several proteins, including anti-oxidant enzymes, notably catalase and SOD and various stress proteins, especially the inducible form of hsp70 (hsp70i). Whole body hyperthermia increases myocardial hsp70i content over a period of several days and myocardial protection can be observed 24 h after hyperthermia. Several lines of evidence suggest that hsp70i is a cytoprotective protein conferring tissue tolerance to ischaemia–reperfusion injury. For example, Hutter *et al.* (1994) showed that the rise in myocardial hsp70i content correlated with both the degree of hyperthermia and the extent of subsequent ischaemic tolerance. Our group showed that the post-hypoxic functional recovery of papillary muscles from rabbits pretreated with hyperthermia correlated with the hsp70i content of an adjacent papillary muscle from the same animal (Marber *et al.*, 1994). More recently, studies with transgenic mice that overexpress hsp70i (Marber *et al.*, 1995; Plumier *et al.*, 1995) and studies involving transfection of the gene encoding for hsp70 into isolated myogenic cells (Heads *et al.*, 1995b; Mestril *et al.*, 1994) and cardiac myocytes (Cumming *et al.*, 1996b) show more convincingly that the protein *directly confers protection* against ischaemic injury. While there is good evidence then that the presence of hsp70 is associated with an ischaemia-tolerant phenotype, the extent to which this particular stress protein accounts for myocardial protection following heat stress is unclear. The different timecourses of hsp70i elevation and myocardial protection following hyperthermia do not support the notion that this protein *alone* is involved in post-hyperthermia protection (Currie *et al.*, 1993; Shipley *et al.*, 1995).

Several reports have suggested that transient ischaemia in whole animals and isolated cells induces hsps. Dillmann's group showed that ischaemia in canine myocardium caused increases in mRNA for hsp70 (Dillmann *et al.*, 1986; Mehta *et al.*, 1988). Knowlton *et al.* (1991) reported that four 5 min coronary occlusions followed by reperfusion in the rabbit led to an increase in hsp70i mRNA and a qualitatively determined (immunohistochemical) increase in protein immunoreactivity 24 h later. Work in our own laboratory, employing both *in vitro* and *in vivo* techniques, has been directed at exploring the complex relationships between the induction of several hsps by ischaemic preconditioning and the subsequent second window of protection. Cumming *et al.* (1996a) have shown that metabolic 'preconditioning' of rat cardiac myocytes (10 mM deoxyglucose, 20 mM lactate) caused no induction of hsp60, a several-fold induction of hsp70i and a two-fold induction of hsp90. This form of preconditioning conferred protection against simulated ischaemia 24 h later. In our group's initial report of the second window of protection, Marber *et al.* (1993) showed that elevation of hsp70i (determined by Western blot analysis) occurred in rabbit myocardium 24 h following hyperthermia or preconditioning and at this time-point protection against infarction was conferred by each treatment. Additionally, ischaemic preconditioning, but not hyperthermia, induced hsp60. However, this particular preconditioning protocol (four 5 min coronary occlusions) may not consistently induce hsp70i. For example, in subsequent studies with the protocol we have observed only modest hsp70i induction in rabbits preconditioned in the conscious state (Yang *et al.*, 1996) and no induction of hsp70i mRNA (unpublished results) or protein (Heads *et al.*, 1995a) in rabbits preconditioned in the anaesthetized state, even though the preconditioning protocol reliably confers a second window of protection. In contrast, Tanaka *et al.* (1994) have reported induction of hsp70i in rabbit myocardium 24 and 48 h after four 5 min coronary occlusions, without observing a second window of protection. It seems then that hsp70i induction following ischaemic preconditioning may not be as robust in this experimental animal model as was initially observed, although we have consistently observed delayed protection against infarction in this model. It may well be the case that subtle alterations in hsp70 regulation may be relevant, such as translocations of both the constitutive and inducible isoforms of the protein to selected cell 'compartments' (which conventional Western blot techniques will not detect). Additionally, the regulation of other hsps in response to ischaemia which have not so far been widely investigated, especially hsp27, may be fundamental to ischaemia-induced delayed myocardial protection and protection induced by other agents. The adenosine A1 receptor agonist, CCPA (2-chloro-*N*-cyclopentyl-adenosine), induced delayed protection against ischaemia (see below) and although this agent does not induce hsp70i, changes in the phosphorylation state and/or translocation of hsp27, as well as induction of manganese-SOD (see below), may follow adenosine receptor activation (Heads *et al.*, 1995a).

4.2. Changes in endogenous anti-oxidant activity

Endogenous anti-oxidant proteins play a fundamental role in maintaining homeostasis in the face of intracellular reactive oxygen species production. During a variety of oxidative stresses in which increased production of protein-damaging free radicals can be detected, cells may adapt biochemically by increasing the activity of enzymes that inactivate free radicals, or modify their reactivity. Increases in endogenous anti-oxidant protein activity, notably catalase and SOD, have been observed in many studies investigating the biology of cellular adaptation to stress. This literature is too large to review here but several studies relating specifically to myocardial adaptation and protection may be relevant. The heat stress response in myocardium following whole body hyperthermia, which confers increased tolerance to ischaemia–reperfusion 24 h later in both isolated and *in vivo* hearts, has been associated in some studies with increases in myocardial catalase activity (Karmazyn *et al.*, 1990; Steare and Yellon, 1993). Other adaptive stimuli such as cytokines and endotoxin derivatives similarly induce delayed protection which has been associated with increases in myocardial catalase (Brown *et al.*, 1989) and SOD activity (Wong and Goeddel, 1988). As far as brief ischaemia specifically is concerned, there is some evidence from *in vivo* animal studies that SOD activity may be elevated 24 h following ischaemic preconditioning. Hoshida *et al.* (1993) described the temporal dynamics of manganese-SOD activity following preconditioning in canine myocardium and described a biphasic pattern of enzyme activity over a 24 h period similar to the pattern of ischaemic tolerance observed in a related study (Kuzuya *et al.*, 1993). Following four 5 min coronary occlusions in the dog, manganese-SOD activity was found to increase in the epicardial risk zone within 1 h, returned to normal within 3 h and was then elevated again 24 h later. The myocardial content of manganese-SOD increased gradually in the ischaemic and non-ischaemic zones over 72 h following ischaemic preconditioning. Twenty-four hours following ischaemic preconditioning, manganese-SOD activity had increased significantly in the subepicardial and subendocardial regions of the ischaemic and non-ischaemic zones but the increase was greatest in the subendocardial region of the ischaemic zone. Interestingly at no point were there any differences in either activity or content of the cytosolic copper-zinc-SOD between the ischaemic and non-ischaemic zones. In a recent preliminary study, we have found that the activities of both manganese-SOD and copper-zinc-SOD are increased 24 h after preconditioning of rabbit myocardium (Heads *et al.*, 1995a). In elegant studies with isolated rat cardiac myocytes, Yamashita *et al.* (1994) reported that hypoxic preconditioning of these cells resulted in increased activity of manganese-SOD 24 h later, at which time-point marked protection was observed against a prolonged hypoxic insult. Both the rise in enzyme activity after preconditioning and the acquisition of cellular tolerance to hypoxia were abolished when myocytes were treated with staurosporine, a protein kinase inhibitor, or with antisense oligonucleotide directed against manganese-SOD during preconditioning. The involvement of a protein kinase mediator in this regulation of SOD activity is particularly interesting to us. We found that treatment of rabbits *in vivo* with the selective adenosine A1 agonist CCPA, conferred protection against infarction 24 h later (Baxter *et al.*, 1994) and was associated with elevated SOD activity (Heads *et al.*, 1995a). A recent report by Maggirwar *et al.* (1995) proposed a link between adenosine receptor activation and increased SOD activity, possibly by both transcriptional and post-translational mechanisms. This association between adenosine, an obviously important paracrine substance in myocardial ischaemia, and SOD would appear to be very relevant to further exploration of the second window of protection.

4.3. Inducible nitric oxide synthase

Parratt and Vegh have suggested that endothelial-derived nitric oxide is an endogenous anti-arrhythmic substance and that classic ischaemic preconditioning against arrythmias in the dog is mediated by release of nitric oxide (Vegh *et al.*, 1992). More recently, they demonstrated that the second window of protection against arrhythmias in this species, induced by rapid ventricular pacing, was abolished by administration of dexamethasone (Vegh *et al.*, 1994). This corticosteroid is appreciated to have many actions, including inhibition of cyclooxygenase and inducible nitric oxide synthase activities. It may also be relevant to this discussion that bacterial endotoxins induce nitric oxide synthase activity and Parratt's group have shown that the delayed anti-arrhythmic effect of endotoxin in the rat heart is abolished by dexamethasone pretreatment (Song *et al.*, 1994). So far, however, there is no direct evidence that nitric oxide synthase activity is increased in myocardium 24 h after ischaemic or other preconditioning stimuli.

4.4. Intracellular signalling cascade: adenosine and protein kinase C

As with classic preconditioning, it seems likely that our understanding of the second window of protection may profit from attention to intracellular signal transduction pathways (see *Figure* 2). In the rabbit, there is a wealth of evidence that the early protection against infarction associated with preconditioning is initiated by adenosine, generated as a consequence of ATP catabolism during the preconditioning ischaemic stimulus (see Chapter 1). With regard to the delayed protection in the rabbit we showed that adenosine receptor blockade with SPT during preconditioning abolished the protection against infarction 24 h later (Baxter *et al.*, 1994). We also demonstrated in the same study that a single bolus of CCPA (25, 50 or 100 $\mu g\ kg^{-1}$) led to enhanced resistance to infarction 24 h later in a rabbit model of coronary occlusion *in vivo*. These observations suggested that adenosine receptor activation during preconditioning in the rabbit might be an important trigger not only of early protection but also of the delayed protection. Interestingly, we have observed a prolonged period of delayed protection following CCPA treatment: an approximately 50% reduction in infarct-to-risk ratio was observed over a period between 24 and 72 h after CCPA treatment but no protection was observed 96 h after treatment (Baxter and Yellon, 1994). The temporal profile of delayed protection following CCPA closely parallels the timecourse of delayed protection following ischaemic preconditioning in the rabbit. The similarity of timecourse and magnitude of protection evoked by ischaemic preconditioning and CCPA pretreatment provide further indirect evidence that in the rabbit adenosine is an important trigger of the second window of protection. We can not exclude the possibility, however, that other locally released

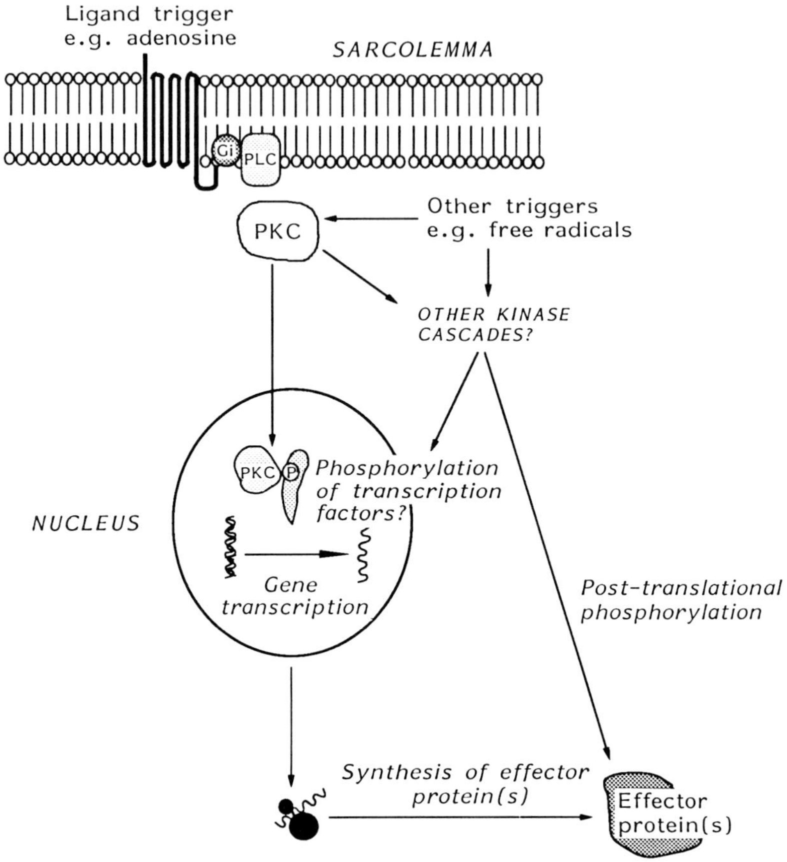

Figure 2. Schematic indicating possible intracellular signalling pathways underlying the second window of protection. Gi, inhibitory GTP-regulatory protein; PKC, protein kinase C; PLC, phospholipase C.

mediators may be involved as triggers of delayed protection in the rabbit. For example, it is known that noradrenaline may be an important trigger for early preconditioning in the rat (Banerjee *et al.*, 1993), and treatment with noradrenaline in the rat can induce a delayed phase of protection in this species that extends to 72 h (Meng *et al.*, 1993).

There is some indirect evidence for the involvement of protein kinase C (PKC) in the second window of protection, since delayed protection was abolished when preconditioning occurred in the presence of chelerythrine, a PKC inhibitor (Baxter *et al.*, 1995b). More recently, we have observed that pretreatment of rabbits *in vivo* with 1,2-dioctanoyl-*sn*-glycerol (a synthetic diacylglycerol activator of PKC) resulted in an approximately 50% reduction in infarct size 24 h later (unpublished results). PKC could influence many intranuclear transcription events, either directly by phosphorylation of transcription factors or indirectly through activation of other kinase signal cascades (e.g. raf kinase and MAP kinase families; Hug and Sarre, 1993). As mentioned above, we have found that 24 h following CCPA pretreatment, activities of both mitochondrial and cytosolic fractions of SOD are increased in rabbit myocardium (Heads *et al.*, 1995a) although changes in myocardial hsp70i content were not observed. However, the complex relationships between protein activity and the protected state following preconditioning and A1 agonist treatment remain to be more fully investigated.

In completing this discussion of the possible mechanisms of the second window of protection we would stress that several forms of delayed protection have been described recently in different species (e.g. against infarction in rabbit and dog, against stunning in the pig, and against arrhythmias in the dog, assessing cell viability in isolated rat cardiac myocytes). Furthermore, the preconditioning or chemical pretreatment protocols in these studies are varied and in some respect may be very particular to the endpoint studied. Two points, which may be self-evident but are nevertheless crucial, arise from these considerations: (i) the cellular mechanisms of protection may vary according to species, model and endpoint; (ii) different preconditioning stimuli may induce diverse patterns of adaptation and protection in the same species. Further investigation of the second window of protection must fully address characterization of experimental models before we are realistically to assess and extrapolate any mechanistic studies.

We hope that an understanding of the physiology of delayed preconditioning will shed light on the associations between antecedent angina pectoris and myocardial infarction (see Chapter 10), and promote the rational development of new therapeutic approaches to cardioprotection in patients with coronary artery disease, especially those about to undergo high-risk elective procedures such as coronary angioplasty and cardiac surgery. As we learn more about its characteristics and mechanisms we may find that delayed preconditioning protection has greater potential for therapeutic exploitation than classic preconditioning. Although its investigation may present as many, if not more, challenges to research than classic preconditioning protection, a fuller appreciation of this innate adapative capacity of myocardium should contribute to our understanding of myocardial responses to ischaemia and the pathophysiology of ischaemic diseases.

References

Banerjee A, Locke-Winter C, Rogers KB, Mitchell MB, Brew EC, Cairns CB, Bensard DD and Harken AH. (1993) Preconditioning against myocardial dysfunction after ischemia and reperfusion by an alpha-adrenergic mechanism. *Circ. Res.* **73,** 656–670.

Baxter GF and Yellon DM. (1994) Temporal characterization of the 'second window of protection': prolonged anti-infarct effect after adenosine A1 receptor activation. *Circulation* **90** (suppl. I), I-475.

Baxter GF, Marber MS, Patel VC and Yellon DM. (1994) Adenosine receptor involvement in a delayed phase of protection 24 hours following ischemic preconditioning. *Circulation* **90,** 2993–3000.

Baxter GF, Goma FM and Yellon DM. (1995a) Involvement of protein kinase C in the delayed cytoprotection following sublethal ischaemia in rabbit myocardium. *Br. J. Pharmacol.* **115,** 222–224.

Baxter GF, Goma FM and Yellon DM. (1995b) Duration of the 'second window of protection' following ischaemic preconditioning in the rabbit. *J. Mol. Cell Cardiol.* **27,** A162.

Brown JM, Grosso MA, Terada LS, Whitman GJ, Banerjee A, White CW, Harken AH and Repine JE. (1989) Endotoxin pretreatment increases myocardial catalase and decreases ischemia–reperfusion injury of isolated rat hearts. *Proc. Natl Acad. Sci. USA* **86,** 2516–2520.

Cohen MV, Yang X-M and Downey JM. (1994) Conscious rabbits become tolerant to multiple episodes of ischemic preconditioning. *Circ. Res.* **74,** 998–1004.

Cumming DVE, Heads RJ, Brand NJ, Yellon DM and Latchman DS. (1996a) The ability of heat stress and metabolic preconditioning to protect primary rat cardiac myocytes. *Basic Res. Cardiol.*, in press.

Cumming DVE, Heads RJ, Watson A, Latchman DS and Yellon DM. (1996b) Differential protection of primary rat cardiocytes by transfection of specific heat stress proteins. *Basic Res. Cardiol.*, in press.

Currie RW, Karmazyn M, Kloc M and Mailer K. (1988) Heat shock response is associated with enhanced post-ischemic ventricular recovery. *Circ. Res.* **63,** 543–549.

Currie RW, Tanguay RM and Kingma JG. (1993) Heat shock response and limitation of tissue necrosis during occlusion–reperfusion in rabbit hearts. *Circulation* **87,** 963–971.

Dillmann WH, Mehta HB, Barrieux A, Guth BD, Neeley WE and Ross J. (1986) Ischemia of the dog heart induces the appearance of a cardiac mRNA coding for a protein with migration characteristics similar to the heat shock/stress protein 71. *Circ. Res.* **59,** 110–114.

Donnelly TJ, Sievers RE, Vissern FLJ, Welch WJ and Wolfe CL. (1992) Heat shock protein induction in rat hearts. A role for improved myocardial salvage after ischemia and reperfusion? *Circulation* **85,** 769–778.

Heads RJ, Baxter GF, Latchman DS and Yellon DM. (1995a) Delayed protection in rabbit heart following ischaemic preconditioning is associated with modulation of HSP27 and superoxide dismutase at 24 h. *J. Mol. Cell Cardiol.* **27,** A163.

Heads RJ, Yellon DM and Latchman DS. (1995b) Differential cytoprotection against heat stress or hypoxia following expression of specific stress protein genes in myogenic cells. *J. Mol. Cell Cardiol.* **27,** 1669–1678.

Hoshida S, Kuzuya T, Fuji H, Yamashita N, Oe H, Hori M, Suzuki K, Taniguchi N and Tada M. (1993) Sublethal ischemia alters myocardial antioxidant activity in canine heart. *Am. J. Physiol.* **264,** H33–H39.

Hug H and Sarre TF. (1993) Protein kinase C isoenzymes: divergence in signal transduction. *Biochem. J.* **291,** 329–343.

Hutter MA, Sievers RE, Barbosa V and Wolfe CL. (1994) Heat shock protein induction in rat hearts: a direct correlation between the amount of heat shock protein induced and the degree of myocardial protection. *Circulation* **89,** 355–360.

Karmazyn M, Mailer K and Currie RW. (1990) Acquisition and decay of heat shock enhanced post-ischemic ventricular recovery. *Am. J. Physiol.* **259,** H424–H431.

Kaszala K, Vegh A, Parratt JR and Papp JG. (1995) Time course of pacing induced preconditioning in dogs. *J. Mol. Cell Cardiol.* **27,** A145.

Knowlton AA, Brecher P and Apstein CS. (1991) Rapid expression of heat shock protein in the rabbit after brief cardiac ischemia. *J. Clin. Invest.* **87**, 139–147.

Kuzuya T, Hoshida S, Yamashita N, Fuji H, Oe H, Hori M, Kamada T and Tada M. (1993) Delayed effects of sublethal ischemia on the acquisition of tolerance to ischemia. *Circ. Res.* **72**, 1293–1299.

Maggirwar SB, Dhanraj DN, Somani SM and Ramukumar V. (1994) Adenosine acts as an endogenous activator of the cellular antioxidant defence mechanism. *Biochem. Biophys. Res. Commun.* **201**, 508–515.

Marber MS, Latchman DS, Walker JM and Yellon DM. (1993) Cardiac stress protein elevation 24 h after brief ischemia or heat stress is associated with resistance to myocardial infarction. *Circulation* **88**, 1264–1272.

Marber MS, Walker JM, Latchman DS and Yellon DM. (1994) Myocardial protection after whole body heat stress in the rabbit is dependent on metabolic substrate and is related to the amount of the inducible 70-kD heat stress protein. *J. Clin. Invest.* **93**, 1087–1094.

Marber MS, Mestril R, Chi S-H, Sayen R, Yellon DM and Dillmann WH. (1995) Overexpression of the rat inducible 70-kD heat stress protein in a transgenic mouse increases the resistance of the heart to ischemic injury. *J. Clin. Invest.* **95**, 1446–1456.

Maxwell MP, Hearse DJ and Yellon DM. (1987) Species variation in the coronary collateral Circulation during regional myocardial ischaemia: a critical determinant of the rate of evolution and extent of myocardial infarction. *Cardiovasc. Res.* **21**, 737–746.

Mehta HB, Popovich BK and Dillman WH. (1988) Ischemia induces changes in the level of mRNAs coding for stress protein 71 and creatine kinase M. *Circ. Res.* **63**, 512–517.

Meng X, Brown JM, Ao L, Mitchell MB, Banerjee AH and Harken AH. (1993) Norepinephrine induces late cardiac protection preceded by oncogene and heat shock protein overexpression. *Circulation* **88** (suppl. I), I-633.

Mestril R, Chi S-H, Sayen MR, O'Reilly K and Dillmann WH. (1994) Expression of inducible stress protein 70 in rat heart myogenic cells confers protection against simulated ischemia induced injury. *J. Clin. Invest.* **93**, 759–767.

Plumier J-CL, Ross BM, Currie RW, Angelidis CE, Kazlaris H, Kollias G and Pagoulatos GN. (1995) Transgenic mice expressing the human heat shock protein 70 have improved post-ischemic myocardial recovery. *J. Clin. Invest.* **95**, 1854–1860.

Qiu Y, Tang X-L, Sun J-Z, Park S-W, Kalya A and Bolli R. (1995) Does the late phase of ischemic preconditioning protect against infarction and reperfusion arrhythmias in conscious pigs. *Circulation* **92** (suppl. I), I-389.

Shipley JB, Qian Y-Z, Levasseur JE and Kukreja RC. (1995) Expression of the stress proteins HSP-27 and HSP-72 in rat hearts does not correlate with ischaemic tolerance after heat shock. *Circulation* **92** (suppl. I), I-654.

Song W, Furman BL and Parratt JR. (1994) Attenuation by dexamethasone of endotoxin protection against ischaemia-induced ventricular arrhythmias. *Br. J. Pharmacol.* **113**, 1083–1084.

Steare SE and Yellon DM. (1993) The protective effect of heat stress against reperfusion arrhythmias in the rat. *J. Mol. Cell Cardiol.* **25**, 1471–1481.

Strasser R, Arras M, Vogt A, Elsasser A, Schwartz ER, Schlepper M and Schaper W. (1994) Preconditioning of porcine myocardium: how much ischemia is required for induction? What is its duration? Is a renewal of effect possible? *Circulation* **90** (suppl.), I-109.

Sun J-Z, Tang X-L, Knowlton AA, Park S-W, Qiu Y and Bolli R. (1995) Late preconditioning against myocardial stunning: an endogenous protective mechanism that confers resistance to postischemic dysfunction 24 h after brief ischemia in conscious pigs. *J. Clin. Invest.* **95**, 388–403.

Tanaka M, Fujiwara H, Yamasaki K, Miyamae M, Yokota R, Hasegawa K, Fujiwara T and Sasyama S. (1994) Ischemic preconditioning elevates cardiac stress protein but does not limit infarct size 24 or 48 h later in rabbits. *Am. J. Physiol.* **267**, H1476–H1482.

Tang X-L, Qiu Y, Sun J-Z, Park S-W, Maldonado C and Bolli R. (1995) Time-course of late preconditioning against myocardial stunning in conscious pigs. *Circulation* **92** (suppl. I), I-388.

Vegh A, Szekeres L and Parratt JR. (1992) Preconditioning of ischaemic myocardium: involvement of L-arginine nitric oxide pathway. *Br. J. Pharmacol.* **107**, 648–652.

Vegh A, Papp JG and Parratt JR. (1994) Prevention by dexamethasone of the marked antiarrhythmic effects of preconditioning induced 20 h after rapid cardiac pacing. *Br. J. Pharmacol.* **113,** 1081–1082.

Walker DM, Pasini E, Kuckukoglu S, Marber MS, Iliodromitis E, Ferrari R and Yellon DM. (1993) Heat stress limits infarct size in the isolated perfused rabbit heart. *Cardiovasc. Res.* **27,** 962–967.

Wong GH and Goeddel DV. (1988) Induction of manganous superoxide dismutase by tumor necrosis factor: possible protective mechanism. *Science* **242,** 941–944.

Yamashita N, Nishida M, Hoshida S, Kuzuya T, Hori M, Taniguchi N, Kamada T and Tada M. (1994) Induction of manganese superoxide dismutase in rat cardiac myocytes increases tolerance to hypoxia 24 hours after preconditioning. *J. Clin. Invest.* **94,** 2193–2199.

Yamashita N, Kuzuya T, Hoshida S, Nishida M, Igarishi J, Hori M, Kamada T and Tada M. (1995) A 'second window of protection' occurs 24 hours after ischemic preconditioning in rat heart. *J. Mol. Cell Cardiol.* **27,** A517.

Yang X-M, Baxter GF, Heads RJ, Yellon DM, Downey JM and Cohen MV. (1996) Infarct limitation in the second window of protection in conscious rabbits. *Cardiovasc. Res.*, in press.

Yellon DM, Pasini E, Cargnoni A, Marber MS, Latchman DS and Ferrari R. (1992) The protective role of heat stress in the ischaemic and reperfused rabbit myocardium. *J. Mol. Cell Cardiol.* **24,** 895–907.

7

Preconditioning in human ventricular myocytes

John S. Ikonomidis, Toshizumi Shirai, Vivek Rao, Ren-Ke Li, Donald A.G. Mickle and Richard D. Weisel,
Departments of Cardiovascular Surgery and Clinical Biochemistry, The Toronto Hospital and The University of Toronto, 200 Elizabeth Street, Toronto, Ontario, Canada M4N 1V7

1. Introduction

In 1986, Murry, Jennings and Reimer made the striking observation that the magnitude of canine myocardial infarction produced by a 40 min circumflex coronary artery occlusion was reduced by 75% when the myocardium at risk had first been subjected to four 5 min coronary occlusions, each separated by 5 min of reperfusion (Murry *et al.*, 1986). This counterintuitive phenomenon named 'ischaemic preconditioning' represents the most dramatic and reproducible known form of myocardial protection from ischaemic injury.

Classically, ischaemic preconditioning is studied in either *in situ* or in isolated working models of regional ischaemia. However, many studies have examined preconditioning under conditions of global ischaemia, either unmodified (Mitchell *et al.*, 1993) or with cardioplegic arrest (Jennings *et al.*, 1991). Preconditioning has been shown to decrease infarct size dramatically in various species, including dogs (Cohen and Downey, 1990; Li *et al.*, 1990; Murry *et al.*, 1986), rabbits (Iwamoto *et al.*, 1991; Thornton *et al.*, 1990), rats (Li and Kloner, 1992), pigs (Flack *et al.*, 1991;

Ischaemia: preconditioning and adaptation, edited by M.S. Marber and D.M. Yellon

Kida *et al.*, 1991; Schott *et al.*, 1990) and horses (Williams *et al.*, 1991) by a variety of investigators. Similarly, the anti-arrhythmic properties of preconditioning have been demonstrated in a number of species (Shiki and Hearse, 1987; Vegh *et al.*, 1990a,b).

The important question with regard to preconditioning is whether it occurs in humans. If preconditioning can be exploited clinically, then elucidation of the mechanisms of this phenomenon may result in new therapeutic strategies in cardiovascular medicine.

Two recent reviews have addressed this issue (Kloner and Yellon, 1994; Lawson, 1994). Four lines of published evidence have developed to support the presence of preconditioning in humans: (i) studies showing that acute tolerance to angina can develop; (ii) clinical studies that suggest that antecedent angina benefits patients suffering from myocardial infarction; (iii) angioplasty studies which demonstrate improvement in various parameters of myocardial ischaemia with successive balloon occlusions; (iv) studies performed on human myocardial tissue specimens taken at operation which show properties of preconditioning. Of the above four lines of evidence, data obtained from studies performed on human myocardial tissue are the most scarce. Yellon and associates randomized 14 patients undergoing elective coronary bypass surgery into two groups (Yellon *et al.*, 1993). While on cardiopulmonary bypass, the preconditioned group received two 3 min episodes of aortic crossclamping with 2 min of reperfusion after each episode, followed by 10 min of sustained crossclamp during which the first saphenous vein conduit was constructed. The control group received only the 10 min crossclamp period without antecedent brief ischaemia. Myocardial biopsies were taken to assess adenosine triphosphate (ATP) levels at baseline, following preconditioning, and following 10 min aortic crossclamping. The results showed that although ATP levels dropped considerably with control ischaemia (from baseline 19.8 to 6.8 $\mu mol\ g^{-1}$ dry weight), no such decrease was observed in the preconditioned patients (from 10.5 following preconditioning to 12.0 $\mu mol\ g^{-1}$ dry weight). The number of patients in this study has subsequently been increased to 10 patients per group with similar results for ATP and trends toward a decrease in lactate production but a statistically significant lowering of CP in the preconditioned group (Alkhulaifi *et al.*, 1994). These studies are very important for two reasons. Firstly, they were the first studies to show a direct biochemical modification by preconditioning in human myocardium. Secondly, these studies show that patients who have already experienced brief myocardial ischaemic episodes (such as with antecedent angina pectoris) may still be preconditioned by an intervention at the time of surgery. Having said this, an exclusion criterion for these studies was unstable angina; thus, this study does not address the issue of whether a similar preconditioning strategy would benefit a higher risk patient subgroup.

Yellon's group (Speechly-Dick *et al.*, 1995; Walker *et al.*, 1993) studied perfused human right atrial trabeculae and found that a 3 min episode of rapid pacing in conjunction with hypoxic perfusion followed by 10 min slow pacing and reoxygenation significantly reduced the deficit in developed tension resulting from 90 min of rapid-paced hypoxia and 120 min slow-paced reoxygenation from 77% to 51% (Walker *et al.*, 1993). In addition they were able to observe that this muscle preparation was able to demonstrate characteristics of the preconditioning phenomenon, that is, it responded to adenosine recepter and ATP-sensitive potassium (K_{ATP}) channel activation in addition to being blocked by antagonists of the adenosine receptor and K_{ATP} channel.

At The Toronto Hospital, we developed the technology to culture human ventricular cardiomyocytes obtained from patients at the time of open heart surgery. We utilized these cultures to document the extent of cell damage following prolonged episodes of simulated ischaemia and reperfusion.

In this chapter, we will review our data which show that cultured human ventricular cardiomyocytes obtained at the time of open heart surgery may be preconditioned with brief episodes of simulated ischaemia. We also provide evidence that this process relies upon release of adenosine by the ischaemic cardiomyocyte, resulting in binding to membrane surface adenosine receptors to initiate protection which is transduced intracellularly, at least in part, by protein kinase C (PKC).

2. Characterization of human ventricular myocyte cultures

Cultures of human ventricular myocytes obtained during cardiac surgery were grown as has been described previously (Li *et al.*, 1992). Ventricular biopsies of 5–20 mg were obtained from tetralogy of Fallot patients undergoing corrective cardiac surgery. Following washing in phosphate-buffered saline (PBS; composition: NaCl 136.9 mM, KCl 2.7 mM, Na_2HPO_4 8.1 mM, KH_2PO_4 1.5 mM, pH 7.3) the connective tissue was removed and the remaining myocardial cells were separated using enzymatic digestion with a mixture of 0.2% trypsin (Difco Laboratories, Detroit, MI, USA) and 0.1% collagenase (Worthington Biochemical Corp., Freehold, NJ, USA). The separated cells were then seeded on to cell culture dishes and cultured at 37°C and 5% CO_2 in Iscove's modified Dulbecco's medium (GIBCO Laboratories, Life Technologies, Grand Island, NY, USA) containing 10% fetal bovine serum, 100 U ml^{-1} penicillin and 100 mg ml^{-1} streptomycin and 0.1 mM β-mercaptoethanol. Purification was achieved using the dilution cloning technique (Freshney, 1983). Using this method, enzymatically isolated cells were seeded at a low density (50–100 cells per 9.0 cm diameter culture dish). With this low seeding, individual cardiomyocytes could be distinguished morphologically by their rectangular shape and large size (40 x 80 μm) from fibroblasts which are smaller and elongated in shape. All cells on the plate were allowed to divide and by 7–14 days, cardiomyocyte and fibroblast colonies were formed. Single cardiomyocyte colonies were then transferred using a Pasteur pipette to another culture dish. Culture purity greater than 95% was demonstrated with fluorescent monoclonal antibody staining for human ventricular myosin heavy chains (Rougier Bio-Tech Ltd, Montreal, Quebec, Canada).

Our characterization of these cells has been extensive. Myocyte homogenates have been shown by Northern blot analysis to contain significant ventricular myosin heavy chain mRNA (Li *et al.*, 1994). In addition, the cells stain positively with human cardiac-specific monoclonal antibody against myoglobin, troponin T, troponin I and the MB isozyme of creatine kinase (CK-MB, unpublished data). The CK-MB activity of these cells is high to passage 9 and is not present in cultured endothelial cells (Li *et al.*, 1992). Electron microscopy of these cells shows abundant mitochondria, extensive but disrupted sarcomeric networks and intercalated discs.

These cardiomyocytes undergo some cellular disorganization and loss of sarcomeres which correlates with increasing time of culture. These changes, which are similar to the changes described by Moses and Claycomb (1989) for adult rat cardiomyocytes, render these cells quiescent. Despite the loss of functional activity, these cells retain many of the characteristics of *in vivo* myocardium. We have recently demonstrated (unpublished data) that the cardiac cell culture antioxidant profiles, ratio of CK to CK-MB and membrane phospholipid and fatty acid composition are similar to these same parameters in human myocardial tissue. Further, when given the choice, these cells preferentially utilize lactate over glucose as a substrate (unpublished data), a phenomenon we previously observed in the myocardium of patients undergoing aorto-coronary bypass surgery. With these cultures we reported differences in water-soluble antioxidant specificity against free radical injury (Mickle *et al.*, 1990) and identified oxygen responsive genetic sequences which regulate glutathione peroxidase expression in response to culture under varying oxygen tensions (Cowan *et al.*, 1992, 1993). Finally, as discussed below, we demonstrated the usefulness of these cells for studies of cellular damage and metabolism following simulated ischaemia (Tumiati *et al.*, 1994).

3. The model of simulated ischaemia and reperfusion

The technique used for simulating cell culture ischaemia and reperfusion was defined previously by Tumiati *et al.* (1994). Following stabilization in PBS with glucose at 37°C for 30 min, ischaemia was simulated by placing the cells into a plexiglass chamber (*Figure 1*) flushed with 100% nitrogen and exposing them to a low volume (1.5 ml) of anoxic (pO_2 = 0 mmHg) perfusion PBS at 37°C for 90 min. The PBS was de-gassed in 200 ml quantities through bubbling with 95% N_2/5% CO_2 that had first been passed through an oxygen trap first of a 1% w/v solution of $NaSO_3$ in de-ionized water (trap no. 1, *Figure 1*) followed by a bicarbonate buffer (Na_2CO_3 20 mM, $NaHCO_3$ 20 mM, trap no. 2, *Figure 1*). The solution was de-gassed until the pO_2 was measured to be 0 mmHg (Blood Gas Analyzer Model IL1312, Instrumentation Laboratory, Milan, Italy). Following this, the gas was changed to 100% N_2 and passed over the perfusion PBS to reduce the pCO_2 to less than 10 mmHg. The solution pH was adjusted to 7.40 ± 0.05 and the osmolality was corrected to 290 ± 10 $mOsm^{-1}$ using 1.0 M NaOH and NaCl, respectively.

In order to verify the desired conditions within the nitrogen chamber, 2 ml of anoxic perfusion PBS was also placed in an open dish within the chamber and tested to ensure the absence of oxygen 5 min from the end of each 'ischaemia' experiment.

'Reperfusion' was accomplished by exposure to 15 ml of normoxic, 37°C perfusion PBS for 30 min.

Using this model, 90 min of 'ischaemia' followed by 30 min of 'reperfusion' was observed to produce approximately 45% cell death (as measured by trypan blue uptake), with associated ATP depletion, accumulation of lactate and development of a significant extracellular acidosis.

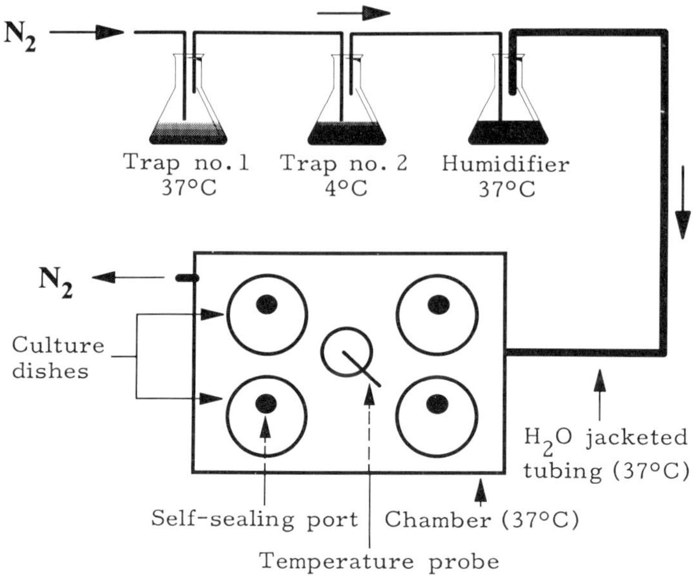

Figure 1. Schematic representation of the simulated ischaemia apparatus. Four plates of cardiomyocytes are sealed within a specially designed plexiglass chamber and exposed to low volume, humidified anoxia. A smaller central dish allows for monitoring of pO_2 and temperature (see text). Reproduced from Ikonomidis *et al.* (1994b) with permission from the Cardiovascular Research Institute, Amsterdam.

4. The human ventricular myocyte response to 'ischaemic' preconditioning

We became interested in whether this model could be used to evaluate ischaemic preconditioning in cultured cells (Ikonomidis *et al.*, 1994a,b,c). We began by assessing whether one cycle of 10 min of 'ischaemia' followed by 10 min 'reperfusion' was effective in protecting against the prolonged 'ischaemia' and reperfusion insult already defined. The results (*Figure 2*) demonstrated that preconditioning resulted in a decrease in trypan blue uptake compared to control 'ischaemic' heart cells. We next tested three preconditioning regimens and showed that 20 min of 'ischaemia' followed by 20 min 'reperfusion' conferred better protection than 10 min of 'ischaemia' followed by 10 min 'reperfusion' (*Figure 3*). The 20 min preconditioning regimen resulted (mean±SEM) in less hydrogen ion (control: 2.01±0.43; preconditioning: 1.40±0.11, μmol mg^{-1} protein, $p < 0.05$) and less LDH release (control: 20 738±3119; preconditioning: 11 927±4227 units mg^{-1} protein, $p < 0.05$). We then

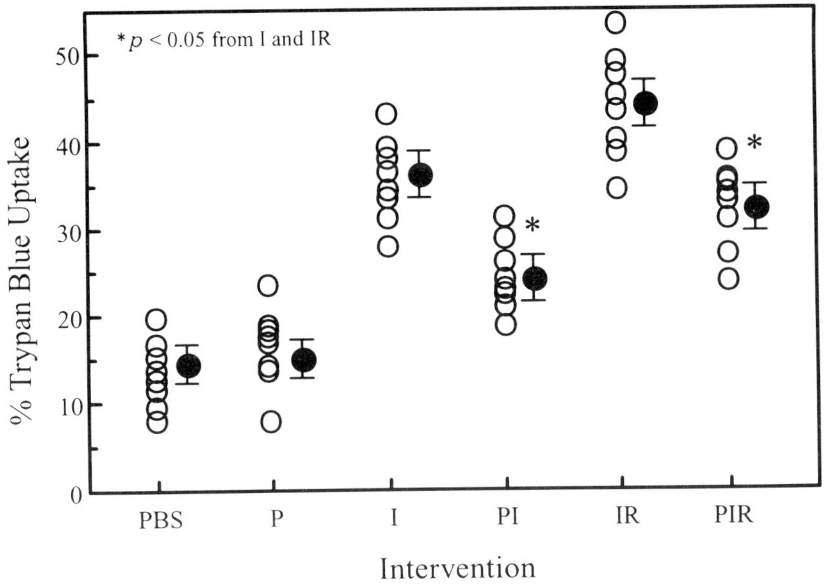

Figure 2. Cell injury with the 10 min 'ischaemia'/10 min 'reperfusion' preconditioning regimen. Preconditioning reduced the percentage of cells which take up trypan blue following 90 min 'ischaemia' with or without 'reperfusion'. PBS 30 min = stabilization in 'perfusion' phosphate-buffered saline (PBS) for 30 min; P = preconditioning; I = 90 min 'ischaemia'; PI = preconditioning plus 'ischaemia'; IR = 'ischaemia' plus 30 min 'reperfusion'; PIR = preconditioning plus 'ischaemia' plus 'reperfusion'. All values are expressed as mean ± SEM, n = eight plates per group. Reproduced from Ikonomidis *et al.* (1994b) with permission from the Cardiovascular Research Institute, Amsterdam.

assessed both intracellular and extracellular lactate release in control cells and cells preconditioned with 20 min of 'ischaemia' followed by 20 min 'reperfusion' at 30, 60 and 90 min of prolonged 'ischaemia'. The results showed that both intracellular and extracellular lactate increased progressively during 'ischaemia' in the control and preconditioned groups; however by 90 min of 'ischaemia', the control group had produced significantly greater lactate (intracellular: control: 1.55±0.62; preconditioned: 0.54±0.23, μmol μg^{-1} DNA, $p < 0.05$; extracellular: control: 0.47±0.09; PC: 0.33±0.07 μmol μg^{-1} DNA, $p < 0.05$). There were no differences in ATP depletion between the two groups.

The application of cell culture models to test strategies of myocardial protection against conditions of simulated ischaemia and hypoxia is relatively new. Ganote and colleagues (1993) used an ischaemic rat cardiomyocyte pellet model to report salutary effects of adenosine and adenosine A_1 receptor agonists against 'ischaemic' damage. In isolated rabbit cardiomyocytes, Armstrong and associates (1994a,b) reported that

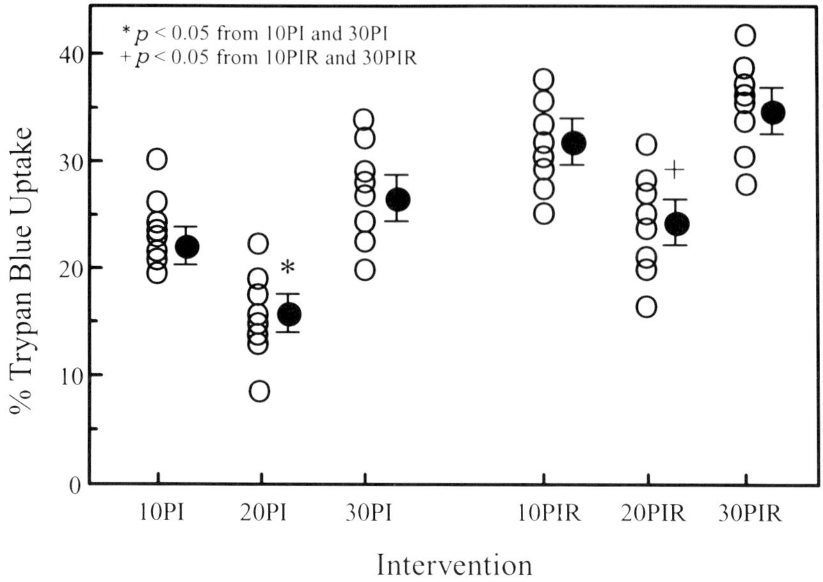

Figure 3. Comparison of different preconditioning times. Preconditioning with 20 min of 'ischaemia' and 20 min of 'reperfusion' conferred the best protection from trypan blue uptake compared to the 10 min and 30 min regimens: 10, 20, 30 PI = preconditioning 'ischaemia' for 10, 20 or 30 min with 10, 20 and 30 min of 'reperfusion' respectively followed by 90 min 'ischaemia'; PIR = preconditioning followed by 'ischaemia' as defined plus 30 min 'reperfusion'. All values are expressed as mean ± SEM, n = eight plates per group). Reproduced from Ikonomidis *et al.* (1994b) with permission from the Cardiovascular Research Institute, Amsterdam.

glucose-free preincubation resulted in a significant preconditioning effect which rivalled that seen in intact heart studies. However, neither of these studies assessed the effects of brief 'ischaemia' and 'reperfusion' for protection against subsequent prolonged 'ischaemia'. Armstrong and Ganote (1994a,b) have subsequently shown that classical 'ischaemic' preconditioning is protective in their rabbit cell model. Cave *et al.* (1993a,b) found that brief hypoxia alone did not produce a preconditioning effect in cultured beating rat cardiomyocytes. The negative outcome of this study differs from our data perhaps because the constant perfusion employed in these studies may have prevented the accumulation of a preconditioning mediator.

In summary, the above data define a unique model of cellular 'ischaemic' preconditioning using a monolayer of cultured human ventricular myocytes and show that preconditioning can occur in cardiomyocytes independent of interactions with other cell types. This model is ideal for the investigation of the potential mechanisms of preconditioning at the cellular level.

5. Mechanisms of preconditioning in human ventricular cardiomyocytes

5.1. The adenosine hypothesis

Ischaemic preconditioning probably begins with the release of some agent or agents from cardiac cells in response to brief ischaemia. This agent then initiates protection against subsequent prolonged ischaemia. This hypothesis is supported by the work of Przyklenk *et al.* (1993) who demonstrated a reduction in infarct size and trends toward preservation of function in the region of myocardium supplied by the left anterior descending (LAD) coronary artery after 1 h of sustained LAD occlusion preceded by a preconditioning stimulus applied to the circumflex coronary artery. Their results suggest that the preconditioning stimulus may be transferred from one region of the heart to another, and that this communication may be humorally based.

Of the many agents endogenous to the myocardium that have been tested for potential preconditioning effects, the most widely studied is adenosine.

Over 65 years ago, the nucleoside adenosine was found to exert widespread effects on a number of organ systems (Drury and Szent-Gyorki, 1929). Adenosine and adenine nucleotides, cellular components which are essential for normal metabolism, were found to direct cellular responses through both intracellular and extracellular mechanisms (Olsson and Pearson, 1990). Under stress conditions, such as ischaemia, intracellular adenosine levels acutely rise as a result of hydrolysis of adenosine monophosphate (AMP) to adenosine through the action of 5′-nucleotidase. Adenosine is released from the cell by passive diffusion (Olsson and Pearson, 1990), and may then serve as a 'retaliatory metabolite' (Newby, 1984) by interaction with specific membrane receptors to effect autocrine or paracrine cellular responses. The result is alteration of a cellular second messenger system to promote restoration of normal cell equilibrium. Interstitial adenosine has a very short (seconds) half-life (Olsson and Pearson, 1990) and is eliminated through specific nucleoside transport or passive diffusion into cells (most notably endothelium) with subsequent deamination to receptor-inactive inosine by adenosine deaminase (Drury and Szent-Gyorki, 1929).

Adenosine is released from cardiomyocytes in large quantities seconds after the onset of ischaemia (Belardinelli *et al.*, 1989). Using microdialysis, Van Wylen (1994) recorded increases in adenosine and other soluble purines in canine myocardial interstitial fluid during the ischaemic and reperfusion phases of preconditioning. Similarly, Kitakaze and colleagues (1993) demonstrated a two-fold rise in coronary venous blood adenosine concentration in preconditioned hearts compared to controls. Inhibition of 5′-nucleotidase with α,β-methylene adenosine 5′-diphosphate (AOPCP) reduced interstitial adenosine and blunted the preconditioning effect (Kitakaze *et al.*, 1994).

Downey and his associates were the first to test adenosine receptor stimulation as a possible initiator of preconditioning. In a rabbit model of preconditioning, two broad acting, dissimilar adenosine receptor blockers, 8-*p*-sulphophenyl theophylline (SPT) and *N*-[2-(dimethyl-amino)ethyl],*N*-methyl-4-(2,3,6,7-tetrahydro-2,6-dioxo-1,3-dipropyl-1H-purin-8-yl)benzosulphonamide (PD 115199), were assessed (Liu *et al.*, 1991). Both were found to block the protective effect of preconditioning on infarction resulting from a 30 min regional coronary artery occlusion. Further, in isolated, blood-perfused hearts, addition of adenosine or the non-selective receptor agonist N^6-(phenyl-2*R*-isopropyl)-adenosine (PIA) resulted in a reduction in infarct size comparable to that seen with preconditioning. Use of PIA in *in situ* heart preparations afforded similar protection to preconditioning whereas adenosine did not, presumably due to the development of profound systemic hypotension. Use of A_2 selective agonists had no effect. The results of this work have since been confirmed by a number of different laboratories, suggesting that adenosine receptor occupation is an important sequel to infarct size reduction after preconditioning in rabbits (Thornton *et al.*, 1992; Tsuchida *et al.*, 1994; Urabe *et al.*, 1993) and dogs (Hoshida *et al.*, 1994; Vander Heide *et al.*, 1993) More recently, adenosine has been shown to exert preconditioning-mimetic effects in human myocardium prior to bypass surgery (Lee *et al.*, 1995). While these results are encouraging, it still remains to be proven that adenosine receptor stimulation can mediate preconditioning in human myocardium.

The effects of adenosine stimulation in rat myocardium may differ from that seen in rabbits. Lasley and colleagues (1993) showed that two A_1 selective adenosine receptor blockers, 8-(4-carboxyethenylphenyl)-1,3-dipropylxanthine (BW A1433U) and 8-cyclopentyl-1,3-dipropylxanthine (DPCPX) were unable to prevent functional recovery following preconditioning prior to global ischaemia. Similar negative results using SPT have been shown by Cave and co-workers (1993a,b). In addition, Asimakis *et al.* (1993) showed that pretreatment with adenosine or PIA may exacerbate postischaemic dysfunction, while BW A1433U did not prevent preconditioning. This discrepancy between rats and rabbits has been suggested to be due to use of crystalloid perfused rather than *in situ* preparations to study rat myocardium; however, preconditioning has also been shown to be effective in crystalloid perfused rabbit hearts (Sandhu *et al.*, 1993). Further, species differences in adenosine metabolism, adenosine receptor number and/or affinity and adenosine coupling mechanisms may account for these differences. There are few data available comparing these variables across different species. However, a recent study comparing human to rat myocardium by Kochan and colleagues (1994) indicates that cellular pathways of intracellular adenosine production and reincorporation are comparable. Another possible reason for the species differences may relate to the different endpoints typically studied with these two models. In the rabbit, the most commonly used endpoint is reduction in infarct size, which appears to be mediated by adenosine receptors. However, in the rat, the primary endpoint is ischaemic dysfunction, which does not appear to be related to adenosine receptor stimulation. One available study in rabbits which assessed functional recovery instead of infarct size showed that SPT pretreatment did not reverse preconditioning-induced improvement in function and further that PIA pretreatment was not protective (Cohen *et al.*, 1991). Further, 3 years prior to the aforementioned adenosine inhibition study performed by Lasley and colleagues (1993), this same group demonstrated that adenosine receptor stimulation prolonged time to ischaemic contracture (more indicative of reduction in cell death than recovery of function) in crystalloid perfused rat hearts (Lasley *et al.*, 1990). In addition, in cultured rat cardiomyocytes, Ganote and colleagues (1993) were able to show protection with adenosine and A_1 selective agonists against cell damage resulting from simulated ischaemia. Hence, the different responses to adenosine seen between rats and rabbits may not reflect species differences, but rather may reflect differences in endpoints measured. These data raise the possibility that while adenosine receptor stimulation may be important for reduction in cellular damage, other mechanisms may be involved in the effects of preconditioning on functional recovery and development of reperfusion arrhythmias.

The above data indicates that, in general, ischaemic preconditioning is dependent upon adenosine A_1 but not A_2 receptor stimulation.

Furthermore recent work has emerged implicating the A_3 receptor in preconditioning. Exposure to N^6-[2-(4-aminophenyl)ethyl]adenosine (APNEA), a compound with approximately four-fold selectivity for A_3 over A_1 receptors, was found to exert a preconditioning-mimetic effect at doses not believed to cause A_1 stimulation in both cultured (Armstrong and Ganote, 1994a) and crystalloid perfused (Liu, G.S. *et al.*, 1994) rabbit myocardium.

All adenosine receptor mediated responses are believed to be transduced through multiple pertussis toxin-sensitive guanosine triphosphate (GTP) binding proteins, known as G proteins (Brown, 1990; Endoh *et al.*, 1993; Munishi and Linden, 1989). In rabbits, pretreatment with pertussis toxin appears to block preconditioning (Thornton *et al.*, 1993a,b). However, perhaps in keeping with the adenosine controversy cited above, there are opposing studies in rats supporting (Lasley and Mentzer, 1993a,b; Piacentini *et al.*, 1993) and refuting (Lawson *et al.*, 1993; Liu and Downey, 1993) involvement of pertussis toxin-sensitive G proteins in preconditioning or adenosine receptor-mediated cardioprotection. However, evidence is increasing that adenosine receptor stimulation may modulate cellular free fatty acid production by phospholipase A2 and inositol phospholipid hydrolysis by phospholipase C (Belardinelli *et al.*, 1989). As discussed below, this activity is associated with initiation of another important signal transduction pathway involving the enzyme PKC.

5.2. The PKC hypothesis

The discovery of PKC was preceded by the development of the second messenger model of cell signalling by Sutherland (1972). The discovery of adenosine 3′,5′-cyclic monophosphate (cAMP) and the model to account for its actions, came after nearly two decades of research aimed at demonstrating that extracellular ligands serve as cofactors for specific intracellular enzymes (Hechter, 1955). The second messenger model prompted researchers to concentrate on events occurring at the cell membrane. The adenylate cyclase system was the first plasma membrane signal transduction mechanism to be characterized. Subsequently, it became evident that the control of the state of phosphorylation of intracellular proteins is an important means by which cellular function is controlled by extracellular messengers (Krebs, 1972).

Receptor-mediated hydrolysis of membrane inositol phospholipids is now realized to be a common mechanism whereby extracellular signals from many hormones, neurotransmitters, antigens, some growth factors and other biologically active substances are transduced into cells. The relationship between stimulation of certain membrane receptors and mobilization of inositol phospholipids was first recognized by Hokin and Hokin (1953), who demonstrated that acetylcholine induces a rapid incorporation of [^{32}P] into phosphatidylinositol and phosphatidic acid. The work of Durell *et al.* (1969) also supported this concept of receptor-mediated phosphatidyl inositide turnover. Subsequently, Michell (1975) postulated that breakdown and resynthesis of inositol phospholipids might open a Ca^{2+} gate. In 1983, Streb and colleagues (1983) showed that inositol-1,4,5-trisphosphate (IP_3), an early product of membrane phosphatidylinositol-4,5-bisphosphate (PIP_2) hydrolysis, is released into the cellular cytoplasm and induces mobilization of Ca^{2+} from an internal store, probably in endoplasmic reticulum. Thereafter, Takai and coworkers (1979) provided evidence that 1,2-diacylglycerol (DAG), the other product of PIP_2 hydrolysis that remains in membranes, initiates the activation of a specialized protein kinase whose signal pathway was separate from, and often synergistic to, the Ca^{2+} signalling pathway for control of various cellular functions. This kinase, called protein kinase C (PKC), was identified by Nishizuka (Takai *et al.*, 1977) and is now known to be ubiquitous. Many aspects of PKC are now well characterized and have been extensively reviewed (Hug and Sarre, 1993; Nishizuka, 1986, 1988, 1992; Rasmussen *et al.*, 1992).

Under resting conditions, PKC is recovered mainly from the cytosol in an inactive form. However, when stimulated, this enzyme usually translocates rapidly to membranes (Kraft and Anderson, 1983), possibly along cytoskeletal structural elements such as microtubules (Limas and Limas, 1983; Liu, Y. *et al.*, 1994), usually in a Ca^{2+}-dependent fashion. This process is associated with binding to activated PKC receptors (RACKS) present on cytoskeletal elements and the cell membrane (Hagmann and Burger, 1992; Mochly-Rosen *et al.*, 1992). This preferential binding to RACKS is thought to confer specificity of PKC isoforms for phosphorylation of different proteins. However, translocation may not be synonymous in all cases with kinase activity of PKC (Rasmussen *et al.*, 1992)

At least 10 members of the PKC family have been identified. While each of these forms are serine/threonine kinases of similar structure and molecular weight (68 000–83 000) with most types being inducible by DAG or phorbol esters, emerging data on the individual family members suggest that these are not merely isozymes of identical function but rather separate enzymes which possess specific tissue and cellular distributions and which are likely to elicit specific cellular responses (Bell and Burns, 1991; Hug and Sarre, 1993; Nishizuka, 1986, 1988, 1992; Rasmussen *et al.*, 1992). In adult myocardium, at least five PKC isozymes have been identified (Kohout and Rogers, 1993) of which the β and ε isoforms seem most abundant (Bogoyevitch *et al.*, 1993; Mochly-Rosen *et al.*, 1992).

It was originally thought that PKC required Ca^{2+} and phospholipid, particularly phosphatidylserine, in conjunction with DAG (elaborated by membrane phospholipase C) for its activation. Newer data show that a later phase of DAG production is mediated by phospholipase D acting on membrane phosphatidylcholine and, further, production of lysophosphatidylcholine and free fatty acids such as arachidonate by phospholipase A_2 may either potentiate or directly stimulate PKC activation (Nishizuka, 1992). Current knowledge has resulted in the division of PKC isoforms into three groups based on differences in structure, activation and modes of activity (Nishizuka, 1992). The Ca^{2+}-dependent or conventional PKCs (cPKCs) of which isoforms α, βI, βII and γ are members, are activated by Ca^{2+}, phosphatidylserine, diacylgycerol and phorbol esters, and this activation is enhanced by *cis* unsaturated fatty acids or lysophosphatidylcholine. The second group is comprised of the four novel PKC (nPKCs) isoforms δ, ε, η(L) and θ. This group is activated by phosphatidylserine, diacylgycerol or phorbol esters, but not Ca^{2+}. Further, the ε subspecies is activated by *cis* unsaturated fatty acids but the δ subspecies is not. Group three is composed of two atypical PKCs (aPKCs), isoforms ζ and λ. These isoforms are affected by phosphatidylserine but not diacylgycerol, phorbol esters or Ca^{2+}. Recent evidence suggests that ζ PKC exhibits a constant low level of activator-independent kinase activity (Nishizuka, 1992).

Recent studies have shown that PKC activation is necessary for ischaemic preconditioning to occur. In rabbits, Brooks and coworkers (1993) demonstrated that MARCKS phosphorylation occurs as early as 5 min into a prolonged episode of myocardial ischaemia, but only in hearts that had first been preconditioned. Ytrehus *et al.* (1994) demonstrated that a single 5 min infusion of the diacylglycerol analogue 1-oleoyl-2-acetylglycerol (OAG) or the phorbol ester 12-*O*-tetradecanoyl-phorbol-13-acetate (PMA) followed by 10 min of washout reduced the infarct size following 30 min of regional myocardial ischaemia by 58 and 77%, respectively. Administration of staurosporine or polymyxin B following preconditioning abolished protection (Ytrehus *et al.*, 1994). Analogous results were obtained by Armstrong and Ganote in an isolated adult rabbit cardiomyocyte model of preconditioning (Armstrong and Ganote, 1994a,b; Armstrong *et al.*, 1994). Further, Y. Liu and colleagues (1994) showed that blockade of PKC translocation during ischaemia with colchicine also prevented preconditioning. In rats, Speechly-Dick and coworkers showed a reduction in infarct size similar to preconditioning with administration of the diacylglycerol analogue 1,2-dioctanoylglycerol (DOG) prior to 45 min of prolonged regional ischaemia. The presence of PKC antagonist chelerythrine after preconditioning but before prolonged ischaemia blocked the protective effect (Speechly-Dick *et al.*, 1994). Mitchell *et al.* (1995) demonstrated membrane translocation of multiple PKC isoforms in rat hearts in response to brief ischaemia using specific fluorescent antibodies.

These results suggest that PKC must be active and associated with membranes during prolonged ischaemia for preconditioning to occur and are consistent with a hypothesis that the preconditioning stimulus functions to induce PKC activation and translocation. The activated, membrane-situated enzyme can then immediately exert protective effects during subsequent prolonged ischaemia. However, this hypothesis is confounded to some extent by other studies which show that activation of PKC exerts detrimental effects which may augment damage resulting from prolonged myocardial ischaemia. Yuan *et al.* (1987) showed that administration of PMA to perfused beating rat hearts resulted in dose-dependent negative changes in both inotropy and chronotropy. Further, Ikeda and colleagues (1988) demonstrated that addition of PMA to hypoxic mouse cardiac cells increased CK release, indicative of potentiation of cardiomyocyte damage. Further studies are required to resolve these issues.

With regard to the adenosine hypothesis of preconditioning, activation of adenosine A_1 receptors in various tissues have been shown to stimulate phospholipase C, the membrane enzyme responsible for conversion of PIP_2 to IP_3 and DAG (Alexander *et al.*, 1990; Haggblad and Fredholm, 1987; Okajima *et al.*, 1989; Schiemann *et al.*, 1991; Schweibert *et al.*, 1992; Spielmann *et al.*, 1992). Further, Kohl and colleagues (1990) observed that administration of PIA to electrically driven guinea pig left atrial and papillary muscle preparations increased IP_3 production with a concomitant fall in PIP_2. These results support the hypothesis that adenosine, released from ischaemic cardiomyocytes, exerts an endogenous form of protection through binding to sarcolemmal adenosine receptors with subsequent activation of PKC.

6. The role of adenosine and PKC in the preconditioning of human ventricular myocytes

This section presents a series of studies using the cardiomyocyte preconditioning model to test two hypotheses:

(i) ischaemic preconditioning results in release from human cardiomyocytes of a mediator (adenosine) which may be used to confer ischaemic protection to non-preconditioned cells.
(ii) Human cardiomyocyte 'ischaemic' or 'adenosine' preconditioning requires activation of PKC.

We began by questioning whether ischaemic preconditioning was mediated by an extracellular ligand (Ikonomidis *et al.*, 1994a,b,c). If this were true and the ligand was stable in solution, then exposure of non-preconditioned cells to 'preconditioning' PBS supernatant removed from cells following 20 min of 'ischaemia' should confer protection against prolonged ischaemic conditions. As shown in *Figure 4*, ischaemic supernatant protects cardiomyocytes in a similar manner to preconditioning. We then questioned whether the presence of adenosine in this supernatant may account for the effect by adding the adenosine receptor antagonist SPT to the supernatant prior to exposure to non-preconditioned cells (Ikonomidis *et al.*, 1994a,b,c). As *Figure 4* shows, the addition of 100 μM SPT to preconditioning supernatant abolished its protective effect. High performance liquid chromatography analysis showed that adenosine was present in the preconditioning supernatant at concentrations which exceed the dissociation constant for the adenosine A_1 receptor. We also showed that preconditioning could be mimicked by exogenous administration of a 50 μM solution of adenosine (*Figure 5*) or 100 μM of adenosine receptor agonist PIA prior to prolonged 'ischaemia' and 'reperfusion' (*Figure 4*).

In addition, we examined the role of PKC in preconditioning in this model (Ikonomidis *et al.*, 1995) As *Figure 5* shows, pretreatment for 20 min with 1 μM PMA resulted in protection similar to that of preconditioning which was abolished in the presence of 200 nM calphostin C, a PKC antagonist. In addition, the protective effect of exogenously administered adenosine was blocked by the presence of 200 nM calphostin C (*Figure 5*)

These studies have led us to make several conclusions about our model. First, preconditioned human ventricular cardiomyocytes secrete a protective mediator(s) into the 'ischaemic' supernatant which is

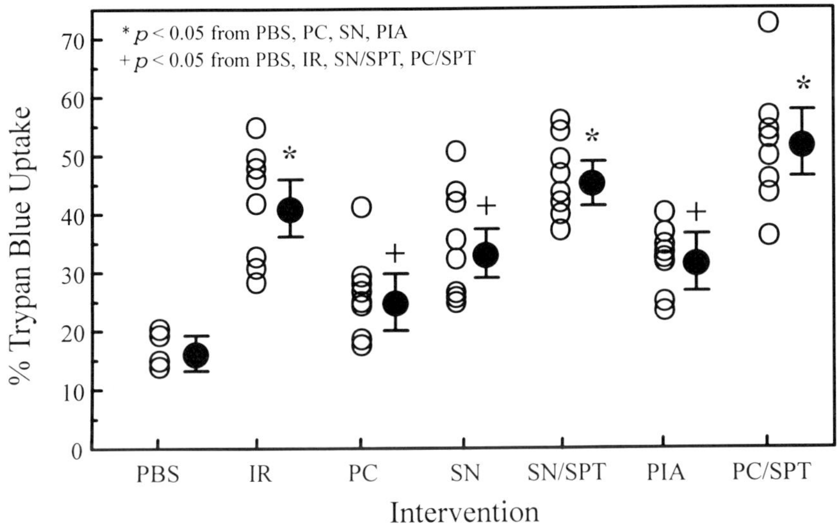

Figure 4. The effect of 'ischaemic' supernatant pretreatment and adenosine receptor agonism or antagonism on cellular damage following prolonged 'ischaemia' and 'reperfusion' (IR) is shown. 'Ischaemic' supernatant pretreatment induced cellular protection similar to preconditioning (PC). Addition of 100 μM 8-*p*-sulphophenyl theophylline (SPT) to the supernatant abolished protection (SN/SPT). Pretreatment with adenosine receptor agonist N^6-(phenyl-2*R*-isopropyl)-adenosine PIA (100 μM) reduced cellular trypan blue uptake similar to PC. Addition of 100 μM SPT prior to and during preconditioning 'ischaemia' abolished protection (SPT/PC). PBS = stabilization in phosphate-buffered saline for 30 min. Open circles represent counts on separate culture plates, while solid circles provide the mean ± SEM.

transferrable to other cells and adenosine is released in sufficient quantities to initiate membrane receptor activation. Second, preconditioning of human ventricular cardiomyocytes is dependent upon PKC. Third, adenosine-mediated protection of human ventricular cardiomyocytes is dependent upon PKC. As summarized in *Figure 6*, these data suggest that adenosine, released from ischaemic human ventricular cardiomyocytes, mediates ischaemic preconditioning through activation of PKC.

7. Possible effector mechanisms of preconditioning

To date, little is known about the effector mechanisms of preconditioning. A number of hypotheses have been tested extensively

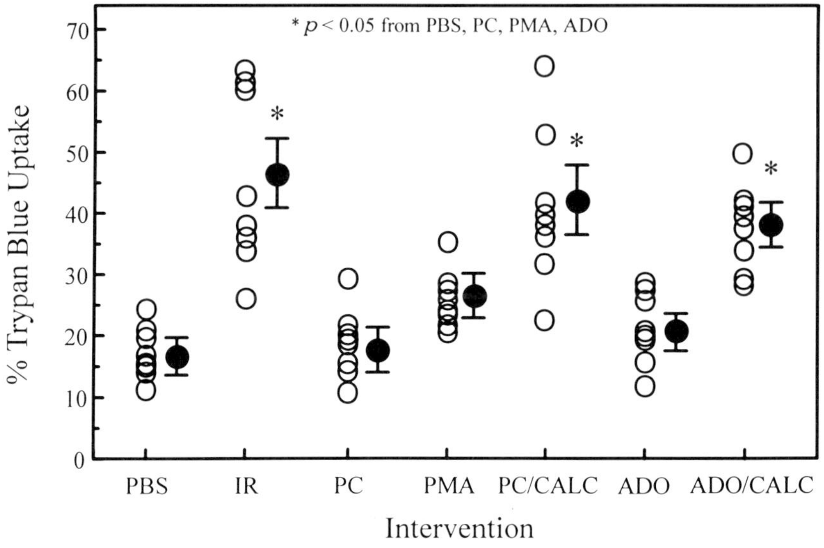

Figure 5. The effect of protein kinase C (PKC) stimulation and inhibition and adenosine stimulation on cellular damage following prolonged 'ischaemia' and 'reperfusion' (IR) is shown. Pretreatment with phorbol ester agonist 12-*O*-tetradecanoyl-phorbol-13 acetate (PMA 1 μM) reduced cellular trypan blue uptake similar to preconditioning (PC) and phosphate-buffered saline stabilization (PBS). Addition of PKC antagonist calphostin C (PC/CALC, 200 nM) prevented preconditioning. Pretreatment with 50 μM adenosine (ADO) reduced cellular trypan blue uptake similar to preconditioning (PC) and PBS stabilization (PBS). Pretreatment with PKC antagonist calphostin C (200 nM) prior to and during adenosine preincubation (ADO/CALC) prevented protection. Open circles represent counts on separate culture plates, while solid circles provide the mean ± SEM. Reproduced from Ikonomidis *et al.* (1994b) with permission from the Cardiovascular Research Institute, Amsterdam.

and are reviewed briefly below. We are currently evaluating some of these possible preconditioning effectors in our human cardiomyocyte model.

7.1 Synthesis of new proteins

One mechanism tested soon after the acceptance of the preconditioning phenomenon was whether brief ischaemia could induce production of a cardioprotective protein in a time period brief enough to coincide with the kinetics of preconditioning. Specifically, the 'heat shock' class of

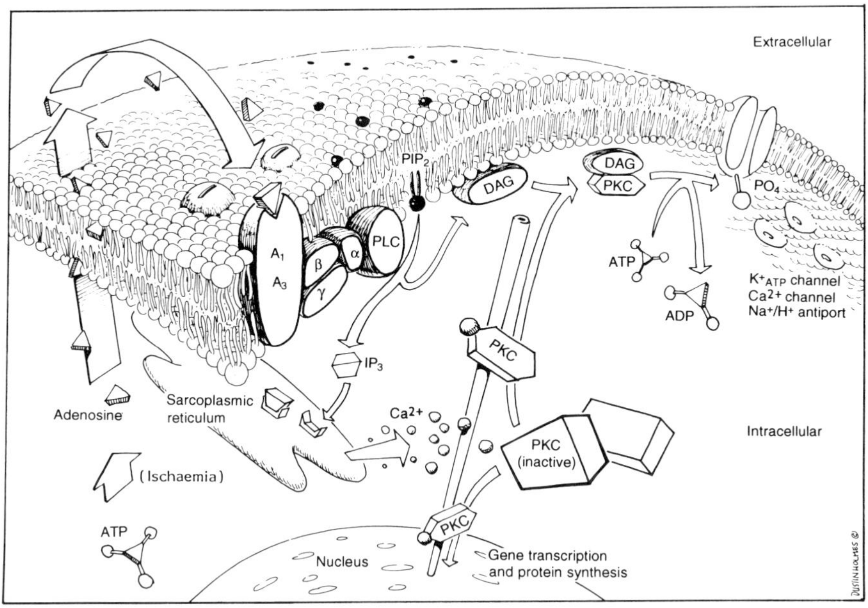

Figure 6. Summary of the adenosine–protein kinase C mechanism of ischaemic preconditioning. Brief ischaemia results in the degradation of adenosine triphosphate (ATP) to adenosine. Adenosine freely diffuses across the cell membrane to interact with surface adenosine receptors. Adenosine receptors are believed to be coupled to inhibitory guanosine triphosphate binding proteins (G_i proteins) consisting of α, B and γ subunits. The activated α subunit stimulates membrane-bound phospholipase C (PLC) to convert membrane phosphatidylinositol bisphosphate (PIP_2) to inositol trisphosphate (IP_3) and diacylglycerol (DAG). IP_3 induces internal mobilization of calcium stores from sites such as the sarcoplasmic reticulum (SR). As the intracellular calcium concentration rises, inactive cytosolic protein kinase C (PKC_{inact}) translocates to cell membranes and is activated by DAG. Activated PKC (PKC_{act}) may now mediate the cardioprotective response through modulation of some final effector. Possibilities for the final effector of acute ischaemic preconditioning include ion channels and exchangers, intermediary metabolic pathways, and in the case of the second window of protection (SWOP) of preconditioning, gene expression and protein synthesis may be activated. Reproduced from Ikonomidis *et al.* (1994b) with permission from the Cardiovascular Research Institute, Amsterdam.

proteins (HSPs), shown to be produced in response to thermal (Currie *et al.*, 1988; Donnelly *et al.*, 1992) and ischaemic (Knowlton *et al.*, 1991a,b) stimuli in a variety of species, were of interest. These proteins are known to be induced through both G protein stimulation (Kiang *et al.*, 1994) and PKC in non-cardiac cells (Faucher *et al.*, 1993; Issandou *et al.*, 1991;

Knowlton *et al.*, 1991a), and thus strengthens our preconditioning signal transduction hypothesis.

This hypothesis has been refuted based on two lines of evidence. First, Das and colleagues (1993) noted that while preconditioning ischaemia resulted in expression in mRNA for HSP 27, HSP 70 and HSP 89, only minimal protein was actually synthesized. Further, neither blockade of DNA transcription to RNA with actinomycin D nor blockade of RNA translation to protein with cycloheximide were found to prevent ischaemic preconditioning (Thornton *et al.*, 1990).

While the acute phase of protection seen with ischaemic preconditioning may not be related to protein synthesis, it is interesting that a 'second window' of protection (SWOP) occurs 24 h after preconditioning (*Figure 6*), and as shown by Marber and coworkers (1993), this response may be related to HSP production.

7.2. Free radicals and other inflammatory mediators

Free radicals. Free radical production following ischaemia is a well-known cause of reperfusion-related damage to the myocardium (Yau *et al.*, 1990). Hence, the hypothesis arose that either a decrease in free radical production or upregulation of antioxidant defenses may be responsible at least in part for preconditioning protection. Both adenosine (Maggirwar *et al.*, 1994) and PKC (Keyser and Alger, 1990) have been implicated in modulation of free radical production.

In rats, Tosaki and colleagues (1994) observed a decrease in malonyldialdehyde (a non-specific indicator of free radical-induced tissue damage) production in preconditioned hearts compared with controls following prolonged ischaemia, probably reflecting a decrease in myocardial damage due to preconditioning. Richard *et al.* (1993) reported that exposing rat myocardium to *N*-2-mercaptopropionyl glycine prior to preconditioning did not prevent the protective effect of preconditioning. The authors concluded from this study that free radicals were not involved in preconditioning. In contrast, Osada and associates (1994) observed that rat hearts preconditioned in the presence of superoxide dismutase (SOD) lost the protective effect of preconditioning on reperfusion arrhythmias.

In rabbits, Tuurens and coworkers (1992) showed that the profiles of several myocardial antioxidant enzymes were unaltered by preconditioning. In addition, Liu and associates (1992) found that removal of neutrophils from the cardiac perfusate did not affect preconditioning. Omar *et al.* (1991) reported that preconditioned rabbit hearts were protected whether or not SOD was present. This result is similar to the findings of Iwamoto and colleagues (1991) who showed no effect of SOD or catalase on preconditioning.

In conclusion, these studies favour a rejection of the free radical hypothesis of preconditioning.

Prostanoids. Another hypothesis tested was whether preconditioning altered the production of prostaglandins and other inflammatory mediators which in turn were responsible for protection. Both adenosine (Bianchi *et al.*, 1991; Neely *et al.*, 1991) and PKC (Ali *et al.*, 1994; Pollard *et al.*, 1993) have been shown to affect these mediator pathways and thus indirectly support this hypothesis.

In rabbits, hearts preconditioned in the presence of cyclooxygenase inhibitors meclofenamate (meclofen) or acetylsalicylic acid (aspirin) still showed a reduction in infarction size compared with controls, suggesting that the cyclooxygenase pathway was not involved in preconditioning (Liu *et al.*, 1991a,b).

In rats, Vegh *et al.* (1990a,b) demonstrated abolition of the protective effect of preconditioning on reperfusion arrhythmias in hearts treated with meclofen. In contrast, Li and Kloner (1992) reported that preconditioning in the rat both with and without aspirin significantly reduced infarct size and the incidence of ventricular arrhythmias.

From the above studies, the influence of preconditioning on prostanoid production is controversial. More studies appear to be required to finalize this issue.

7.3. Intermediary metabolism

In the light of the available data which show that preconditioned myocardium produces less lactate (Ikonomidis *et al.*, 1994a,b,c; Murry *et al.*, 1990) and proteins (Ikonomidis *et al.*, 1994a,b,c), a hypothesis has arisen that preconditioning may induce protection through a reduction in anaerobic glycolysis, thus preventing extreme intracellular acidosis. With regard to the adenosine hypothesis of preconditioning, the effects of adenosine on glucose uptake and metabolism tend to be inconsistent with the findings seen in preconditioning. Adenosine receptor stimulation appears to increase glucose uptake in a variety of tissues including myocardium (Law and Raymond, 1988; Mainwaring *et al.*, 1988; Tominaga *et al.*, 1992; Vergauwen *et al.*, 1994). In terms of glucose metabolism, conflicting reports exist showing an increase (Lasley and Mentzer, 1993a,b; Wyatt *et al.*, 1989) or decrease (Vander Heide *et al.*, 1993) in myocardial lactate production with adenosine receptor stimulation. Also, PKC has been shown to modulate both glucose transport (Kirsch *et al.*, 1985) and the activities of glycogen synthase (Williamson *et al.*, 1985) and phosphofructokinase (Hofer *et al.*, 1985).

In rabbits, promotion of intermediary metabolism through addition of pyruvate to the cardiac perfusate was not found to influence preconditioning (Liu *et al.*, 1992). Further, in two separate studies, Armstrong and associates demonstrated first that glucose-free preincubation (which would inhibit glycolysis) protected against simulated ischaemia (Armstrong *et al.*, 1994), and subsequently that glycolytic inhibition induced by iodoacetate did not block preconditioning in the same model (Armstrong and Ganote, 1994a,b). These results suggest that preconditioning is not mediated by changes in glucose transport or glycolysis in rabbits.

In rats, Wolfe and coworkers reported that preconditioning was associated with a significant reduction in myocardial glycogen stores and attenuation of cellular acidosis compared to controls. Further, in rats that were allowed to recover from preconditioning prior to prolonged ischaemia, the time course of loss of glycogen paralleled the loss of protection from ischaemic injury (Wolfe *et al.*, 1993). A [^{32}P] nuclear magnetic resonance study performed by de Albuquerque *et al.* (1994) showed that both preconditioning and metabolic inhibition with cyanide resulted in preservation in intracellular pH (indicative of a reduction in glycolysis) compared to controls, which correlated with improved functional recovery. Finally, Murphy and coworkers (1993) showed that the adenosine A_1 receptor antagonist BW-A1433U blocked both preconditioning and cellular hexose uptake. These results suggest that inhibition of glycolytic rate is important for preconditioning in rats.

More studies are required to settle the effects of adenosine and PKC on intermediary metabolism and to explain the species differences in the dependence of preconditioning upon glycolysis.

7.4. ATP-sensitive potassium channels

Gross and Auchampach (1992) were the first to propose that opening of K_{ATP} channels mediates preconditioning, an observation supported by other authors. This is an attractive theory because: (i) opening of potassium channels can be protective to the ischaemic heart (Cole *et al.*,

1991); (ii) adenosine and A_1 selective analogues open K_{ATP} channels in rat (Kirsch *et al.*, 1990) and rabbit (Nakhostine and Lamontagne, 1993) heart models and blockade of K_{ATP} channels abolishes adenosine-induced cardioprotection (Van Winkle *et al.*, 1994); (iii) PKC activation is also known to activate K_{ATP} channels in pig myocardium (Black *et al.*, 1993).

In dogs, blockade of the K_{ATP} channel will block the reduction in infarct size seen with preconditioning (Auchampach *et al.*, 1992; Gross and Auchampach, 1992). In addition, pretreatment with K_{ATP} channel blockers has been shown to abolish the protection seen with adenosine receptor stimulation in this species (Auchampach and Gross, 1993; Grover *et al.*, 1992). However, Vegh and colleagues (1990a) were unable to prevent the protective effect of preconditioning on reperfusion arrhythmias with K_{ATP} channel blockade. Overall, these canine studies support the K_{ATP} channel hypothesis of preconditioning.

In rabbits, Thornton and colleagues (1993a,b) demonstrated that K_{ATP} blockade decreases tolerance to ischaemia but does not prevent the infarction-sparing effect of preconditioning. In contrast, Toombs *et al.* (1993) reported the opposite result in a virtually identical model. Further, Tan *et al.* (1993) showed that K_{ATP} blockade prevented the delay in cellular electrical uncoupling seen with preconditioning. The K_{ATP} channel hypothesis is therefore unresolved in rabbits.

In rats, the message is more clear. Liu and Downey (1992) first reported that K_{ATP} blockade did not prevent limitation of infarction due to preconditioning. Lu and coworkers (1993) similarly showed that K_{ATP} blockade did not prevent preconditioning-induced reduction in arrhythmias. Further, K_{ATP} blockade had no effect on preconditioning-induced decreases in LDH release (Grover *et al.*, 1993) or sparing of functional recovery (Cerbai *et al.*, 1988; Fralix *et al.*, 1993). The evidence is therefore convincing that K_{ATP} channels are not involved in preconditioning in rats.

In summary, there are species and outcome measure discrepancies regarding involvement of the ATP-sensitive K^+ channel in the mechanism of preconditioning. Hence, further investigation into this hypothesis is required.

7.5. Modulation of cellular calcium flux

It is known that G_i proteins are coupled to intracellular adenylate cyclase. Stimulation of the G_i protein via the adenosine receptor reduces the activity of this enzyme, resulting in a fall in intracellular cAMP concentrations. cAMP is known to mediate the protein kinase-dependent phosphorylation of the L calcium channel, which causes calcium flux into the cell (Ikonomidis *et al.*, 1990). Calcium has long been known to be a final effector of ischaemic myocardial damage (Buckberg, 1993; Steenbergen *et al.*, 1993a,b). Therefore, a reduction in calcium influx during myocardial ischaemia through inhibition of the L channel may provide the effector mechanism of preconditioning. This hypothesis is attractive because adenosine has been shown to inhibit inward calcium currents from the L channel (Eckert *et al.*, 1993; Kato *et al.*, 1990) and PKC activation has been shown to reduce intracellular calcium availability (Gwathmey and Hajjar, 1991).

To date, there is little published work on the calcium hypothesis. Steenbergen and colleagues (Steenbergen *et al.*, 1993a,b) measured intracellular Ca^{2+} concentrations in ischaemic rat hearts using [^{32}P] nuclear magnetic resonance. This study showed a reduction in intracellular Ca^{2+} concentration in preconditioned hearts compared to control. In pigs, Stokke and associates (1994) did not find any significant differences in L-channel density in preconditioned and control regions of the heart, suggesting that a reduction in calcium influx seen with preconditioning may be due to reduced channel activity or an increase in cellular calcium clearance.

8. Conclusions

Traditionally, cardiac surgeons have attempted to minimize myocardial ischaemic and reperfusion injury during cardiac procedures by optimizing cardioplegic solutions and modifying the conditions of reperfusion. New evidence suggests that in addition to these two strategies, surgeons may be able to increase endogenous myocardial resistance to ischaemic injury through ischaemic preconditioning. Elucidation of the mechanism(s) of this phenomenon may permit the development of a pharmacological agent that induces the preconditioning effect. The clinical application of preconditioning may provide the cardiac surgeon with perhaps the most powerful tool in the armamentarium of myocardial protective techniques which could help meet the challenge of performing operative procedures on increasingly higher risk patients (Ikonomidis *et al.*, 1994a,b,c).

Promising mechanisms arising from studies in cell cultures such as those outlined above should then be tested in a large animal model of myocardial ischaemia and reperfusion, followed by trials in low risk patients undergoing cardiac surgery. Hopefully, the end result of these endeavours will be the development of a new cardioprotective strategy that can be applied clinically.

Acknowledgements

The authors' research has been supported by a grant from the Heart and Stroke Foundation of Canada and by Research Fellowships (to J.S.I.) from the Medical Research Council of Canada and the Heart and Stroke Foundation of Canada.

References

Alexander SPH, Hill SJ and Kendall DA. (1990) Is the adenosine receptor modulation of histamine-induced accumulation of inositol phosphates in cerebral cortical slices mediated by effects on calcium ion fluxes? *J. Neurochem.* **55,** 1138–1141.

Ali A, Ford-Hutchinson AW and Nicholson DW. (1994) Activation of protein kinase C down-regulates leukotriene C4 synthase activity and attenuates cysteinyl leukotriene production in an eosinophilic substrain of HL-60 cells. *J. Immunol.* **153,** 776–788.

Alkhulaifi AM, Yellon DM and Pugsley WB. (1994) Preconditioning the human heart during aorto-coronary bypass surgery. *Eur. J. Cardiothorac. Surg.* **8,** 270–276.

Armstrong S and Ganote CE. (1994a) Adenosine receptor specificity in preconditioning of isolated rabbit cardiomyocytes: evidence of A_3 receptor involvement. *Cardiovasc. Res.* **28,** 1049–1056.

Armstrong S and Ganote CE. (1994b) Preconditioning of isolated rabbit cardiomyocytes: effects of glycolytic blockade, phorbol esters, and ischaemia. *Cardiovasc. Res.* **28,** 1700–1706.

Armstrong S, Downey JM and Ganote CE. (1994) Preconditioning of isolated rabbit cardiomyocytes, induction by metabolic stress and blockade by the adenosine antagonist SPT and calphostin C, a protein kinase C inhibitor. *Cardiovasc. Res.* **28,** 72–77.

Asimakis GK, Inners-McBride K and Conti VR. (1993) Attenuation of postischaemic dysfunction by ischaemic preconditioning is not mediated by adenosine in the isolated rat heart. *Cardiovasc. Res.* **27,** 1522–1530.

Auchampach JA and Gross GJ. (1993) Adenosine A_1 receptors, K_{ATP} channels, and ischemic preconditioning in dogs. *Am. J. Physiol.* **264,** H1327–H1336.

Auchampach JA, Grover GJ and Gross GJ. (1992) Blockade of ischaemic preconditioning in dogs by the novel ATP dependent potassium channel antagonist sodium 5-hydroxydecanoate. *Cardiovasc. Res.* **26,** 1054–1062.

Belardinelli L, Linden J and Berne R. (1989) The cardiac effects of adenosine. *Prog. Cardiovasc. Dis.* **32,** 73–97.

Bell RM and Burns DJ. (1991) Lipid activation of protein kinase C. *J. Biol. Chem.* **266,** 4661–4664.

Bianchi G, Calvani AB, Mantovani M and Prino G. (1991) Enhancement by defibrotide of prostanoid neosynthesis from arachidonic acid: an adenosine receptor-mediated effect? *Sem. Thromb. Hemost.* **17,** 399–403.

Black SC, Fagbemi SO, Chi L, Friedrichs GS and Lucchesi BR. (1993) Phorbol ester-induced ventricular fibrillation in the Langendorff-perfused rabbit heart: antagonism by staurosporine and glibenclamide. *J. Mol. Cell. Cardiol.* **25,** 1427–1438.

Bogoyevitch MA, Parker PJ and Sugden PH. (1993) Characterization of protein kinase C isotype expression in adult rat heart. Protein kinase C-ε is a major isotype present and it is activated by phorbol esters, epinephrine, and endothelin. *Circ. Res.* **72,** 757–767.

Brooks G, Walsh R and Downey JM. (1993) Phosphorylation of 80K/MARCKS, a specific substrate of protein kinase C, occurs after 5′ of ischemia but only in preconditioned hearts (Abstract). *Circulation* **88,** I-191.

Brown AM. (1990) Regulation of heartbeat of G protein-coupled ion channels. *Am. J. Physiol.* **259,** H1621–H1628.

Buckberg GD. (1993) Protean causes of myocardial stunning in infants and adults. *J. Card. Surg.* **8**(suppl. 2), 214–219.

Cave AC, Adrian S, Apstein CS and Silverman HS. (1993a) Anoxic preconditioning fails to reduce subsequent anoxic injury in isolated rat cardiac myocytes (Abstract). *Circulation* **88,** I-544.

Cave AC, Collis CS, Downey JM and Hearse JM. (1993b) Improved functional recovery by ischaemic preconditioning is not mediated by adenosine in the globally ischaemic rat heart. *Cardiovasc. Res.* **27,** 663–668.

Cerbai E, Klockner U and Isenberg G. (1988) Ca-antagonistic effects of adenosine in guinea pig atrial cells. *Am. J. Physiol.* **255,** H872–H878.

Cohen MV and Downey JM. (1990) Myocardial stunning in dogs: preconditioning effect and influence of coronary blood flow. *Am. Heart J.* **120,** 282–291.

Cohen MV, Liu GS and Downey JM. (1991) Preconditioning causes improved wall motion as well as smaller infarcts after transient coronary occlusion in rabbits. *Circulation* **84,** 341–349.

Cole WC, McPherson CD and Sontag D. (1991) ATP-regulated K^+ channels protect the myocardium against ischemic/reperfusion damage. *Circ. Res.* **69,** 571–581.

Cowan DB, Weisel RD, Williams WG and Mickle DAG. (1992) The regulation of glutathione peroxidase gene expression by oxygen tension in cultured human cardiomyocytes. *J. Mol. Cell. Cardiol.* **24,** 423–433.

Cowan DB, Weisel RD, Williams WG and Mickle DAG. (1993) Identification of oxygen responsive elements in the 5′-flanking region of the human glutathione peroxidase gene. *J. Biol. Chem.* **268,** 26904–26910.

Currie RW, Karmazyn M, Kloc M and Mailer K. (1988) Heat-shock response is associated with enhanced postischemic ventricular recovery. *Circ. Res.* **63,** 543–549.

Das DK, Engelman RM and Kimura Y. (1993) Molecular adaptation of cellular defences following preconditioning of the heart by repeated ischemia. *Cardiovasc. Res.* **27,** 578–584.

de Albuquerque CP, Gerstenblith G and Weiss RG. (1994) Importance of metabolic inhibition and cellular pH in mediating preconditioning contractile and metabolic effects in rat hearts. *Circ. Res.* **74,** 139–150.

Donnelly TJ, Sievers RE, Vissern FLJ, Welch WJ and Wolfe CL. (1992) Heat shock protein induction in rat hearts. A role for improved myocardial salvage after ischemia and reperfusion? *Circulation* **85,** 769–778.

Drury AN and Szent-Gyorki A. (1929) The physiological activity of adenine compounds with especial reference to their action upon the mammalian heart. *J. Physiol.* **68,** 213–237.

Durell J, Garland JT and Freidel RO. (1969) Acetylcholine action: biochemical aspects. *Science* **165,** 862–866.

Eckert R, Utz J, Trautwein W and Mentzer RM Jr. (1993) Involvement of intracellular Ca^{2+} release mechanism in adenosine-induced cardiac Ca^{2+} current inhibition. *Surgery* **114,** 334–342.

Endoh M, Takanashi M, Norota I, Kawabata Y and Asano T. (1993) Pronounced direct inhibitory action mediated by adenosine A_1 receptor and pertussis toxin-sensitive G protein on the ferret ventricular myocardium. *Naunyn-Schiedebergs Arch. Pharmacol.* **348,** 282–289.

Faucher C, Capdevielle J, Canal I, Ferrara P, Mazarguil H, McGuire WL and Darbon JM. (1993) The 28-kDa protein whose phosphorylation is induced by protein kinase C activators in MCF-7 cells belongs to the family of low molecular mass heat shock proteins and is the estrogen-regulated 24-kDa protein. *J. Biol. Chem.* **268,** 15 168–15 173.

Flack JE III, Kimura Y, Engelman RM, Rousou JA, Iyengar J, Jones R and Das D. (1991) Preconditioning the heart by repeated stunning improves myocardial salvage. *Circulation* **84,** III-369–III-374.

Fralix TA, Steenbergen C, London RE and Murphy E. (1993) Glibenclamide does not abolish the protective effect of preconditioning on stunning in the isolated perfused rat heart. *Cardiovasc. Res.* **27,** 630–637.

Freshney RI. (1983) *Culture of Animal Cells: a Manual of Basic Technique.* Alan R. Liss, New York, pp. 129–143.

Ganote CE, Armstrong S and Downey JM. (1993) Adenosine and A_1-selective agonists offer minimal protection against ischemic injury to isolated rat cardiomyocytes. *Cardiovasc. Res.* **27,** 1670–1676.

Gross GJ and Auchampach JA. (1992) Blockade of ATP-sensitive potassium channels prevents myocardial preconditioning in dogs. *Circ. Res.* **70,** 223–233.

Grover GJ, Sleph PG and Dzwonczyk S. (1992) Role of myocardial ATP-sensitive potassium channels in mediating preconditioning in the dog heart and their possible interaction with adenosine A_1 receptors. *Circulation* **86,** 1310–1316.

Grover GJ, Dzwonczyk S, Sleph PG and Sargent CA. (1993) The ATP-sensitive potassium channel blocker glibenclamide (glyburide) does not abolish preconditioning in isolated ischemic rat hearts. *J. Pharmacol. Exp. Ther.* **265,** 559–564.

Gwathmey JK and Hajjar RJ. (1991) Protein kinase C activation in human ventricular myocardium. *Biorheology* **28,** 151–160.

Häggblad J and Fredholm BB. (1987) Adenosine and neuropeptide Y enhance α_1-adrenoceptor-induced accumulation of inositol phosphates and attenuate forskolin-induced accumulation of cyclic AMP in rat vas deferens. *Neurosci. Lett.* **82,** 211–216.

Hagmann J and Burger MM. (1992) Phosphorylation of vinculin in human platelets spreading on a solid surface. *J. Cell. Biochem.* **50,** 237–244.

Hechter O. (1955) Concerning possible mechanisms of hormone action. *Vitamin Horm.* **13,** 293–346.

Hofer HW, Schlatter S and Graefe M. (1985) Phosphorylation of phosphofructokinase by protein kinase C changes the allosteric properties of the enzyme. *Biochem. Biophys. Res. Commun.* **129,** 892–897.

Hokin MR and Hokin LE. (1953) Enzyme secretion and the incorporation of ^{32}P into phospholipids of pancreas slices. *J. Biol. Chem.* **203,** 967–977.

Hoshida S, Kuzuya T, Nishida M, Yamashita N, Oe H, Hori M, Kamada T and Tada M. (1994) Adenosine blockade during reperfusion reverses the infarct-limiting effect in preconditioned canine hearts. *Cardiovasc. Res.* **28,** 1083–1088.

Hug H and Sarre TF. (1993) Protein kinase C isoenzymes: divergence in signal transduction? *Biochem. J.* **291,** 329–343.

Ikeda U, Arisaka H, Takayasu T, Takeda K, Natsume T and Hosoda S. (1988) Protein kinase C activation aggravates hypoxic myocardial injury by stimulating Na^+/H^+ exchange. *J. Mol. Cell. Cardiol.* **20,** 493–500.

Ikonomidis JS, Salerno TA and Wittnich C. (1990) Calcium and the heart: an essential partnership. *Can. J. Cardiol.* **6,** 305–316.

Ikonomidis JS, Shirai T, Weisel RD and Mickle DAG. (1994a) Human cardiomyocyte preconditioning is adenosine A_1 receptor dependent. *Circulation* **90,** I-477.

Ikonomidis JS, Tumiati LC, Weisel RD, Mickle DAG and Li R-K. (1994b) Preconditioning human ventricular cardiomyocytes with brief periods of simulated ischemia. *Cardiovasc. Res.* **28,** 1285–1291.

Ikonomidis JS, Weisel RD and Mickle DAG. (1994c) Ischemic preconditioning: cardioprotection for cardiac surgery. *J. Card. Surg.* **9** (suppl.), 526–531.

Ikonomidis JS, Shirai T, Weisel RD, Derylo B, Rao V, Whiteside CI, Mickle DAG and Li R-K. (1995) 'Ischemic' or adenosine preconditioning of human ventricular cardiomyocytes is protein kinase C dependent. *Circulation* **92,** I-12.

Issandou M and Darbon JM. (1991) Activation of protein kinase C by phorbol esters induces DNA synthesis and protein phosphorylation in glomerular mesangial cells. *FEBS Lett.* **281,** 196–200.

Iwamoto T, Miura T, Adachi T, Noto T, Ogawa T, Tsuchida A and Iimura O. (1991) Myocardial infarct size-limiting effect of ischemic preconditioning was not attenuated by oxygen free-radical scavengers in the rabbit. *Circulation* **83,** 1015–1022.

Jennings RB, Murry CE and Reimer KA. (1991) Energy metabolism in preconditioned and control myocardium: effect of total ischemia. *J. Mol. Cell. Cardiol.* **23,** 1449–1458.

Kato M, Yamaguchi H and Ochi R. (1990) Mechanism of adenosine-induced inhibition of calcium current in guinea pig ventricular cells. *Circ. Res.* **67,** 1134–1141.

Keyser DO and Alger BE. (1990) Arachidonic acid modulates calcium current via protein kinase C and oxygen radicals. *Neuron* **5,** 545–553.

Kiang JG, Carr FE, Burns MR and McClain DE. (1994) HSP-72 synthesis is promoted by increase in $[Ca^{2+}]_i$ or activation of G proteins but not Ph_i or CAMP. *Am. J. Physiol.* **267,** C104–C114.

Kida M, Fujiwara H, Ishida M, Kawai C, Ohura M, Miura I and Yabuuchi Y. (1991) Ischemic preconditioning preserves creatine phosphate and intracellular pH. *Circulation* **84,** 2495–2503.

Kirsch C, Obermaier B and Haring HU. (1985) Phorbol esters enhance basal D-glucose transport but inhibit insulin stimulation of D-glucose transport and insulin binding in isolated rat adipocytes. *Biochem. Biophys. Res. Commun.* **128,** 824–832.

Kirsch GE, Codina J, Birnbaumer L and Brown AM. (1990) Coupling of ATP-sensitive K^+ channels to A_1 receptors by G proteins in rat ventricular myocytes. *Am. J. Physiol.* **259,** H820–H826.

Kitakaze M, Hori M, Takashima S, Sato H, Inoue M and Kamada T. (1993) Ischemic preconditioning increases adenosine release and 5′-nucleotidase activity during myocardial ischemia and reperfusion in dogs. Implications for myocardial salvage. *Circulation* **87,** 208–215.

Kitakaze M, Hori M, Morioka T, Minamino T, Takashima S, Sato H, Shinozaki Y, Chujo M, Mori H and Inoue M. (1994) Infarct size limiting effect of ischemic preconditioning is blunted by inhibition of 5′-nucleotidase activity and attenuation of adenosine release. *Circulation* **89,** 1237–1246.

Kloner RA and Yellon DM. (1994) Does ischemic preconditioning occur in patients? *J. Am. Coll. Cardiol.* **24,** 1133–1142.

Knowlton A, Brecher P and Apstein CS. (1991a) Rapid expression of heat shock protein in the rabbit after brief cardiac ischemia. *J. Clin. Invest.* **87,** 139–147.

Knowlton AA, Eberli FR, Brecher P, Romo GM, Owen and Apstein CS. (1991b) A single myocardial stretch or decreased systolic fibre shortening stimulates the expression of heat shock protein 70 in the isolated, erythrocyte-perfused rabbit heart. *J. Clin. Invest.* **88,** 2018–2025.

Kochan Z, Smolenski RT, Yacoub MH and Seymour AL. (1994) Nucleotide and adenosine metabolism in different cell types of human and rat heart. *J. Mol. Cell. Cardiol.* **26,** 1497–1503.

Kohl C, Linck B, Schmitz W, Scholz H, Scholz J and Tóth M. (1990) Effects of carbachol and (-)-N^6-phenylisopropyladenosine on myocardial inositol phosphate content and force of contraction. *Br. J. Pharmacol.* **101,** 829–834.

Kohout TA and Rogers TR. (1993) Use of a PCR-based method to characterize protein kinase C isoform expression in cardiac cells. *Am. J. Physiol.* **264,** C1350–C1359.

Kraft AS and Anderson WB. (1983) Phorbol esters increase the amount of Ca^{2+}, phospholipid-dependent protein kinase associated with plasma membrane. *Nature* **301,** 621–623.

Krebs EG. (1972) Protein kinases. *Curr. Topics Cell. Reg.* **5,** 99–133.

Lasley RD and Mentzer RM Jr. (1993a) Pertussis toxin blocks adenosine A_1 mediated protection of the ischemic rat heart. *J. Mol. Cell. Cardiol.* **25,** 815–821.

Lasley RD and Mentzer RM Jr. (1993b) Adenosine increases lactate release and delays the onset of contracture during global low flow ischemia. *Cardiovasc. Res.* **27,** 1352–1357.

Lasley RD, Rhee JW, Van Wylen DG and Mentzer RM Jr. (1990) Adenosine A_1 receptor mediated protection of the globally ischemic isolated rat heart. *J. Mol. Cell. Cardiol.* **22,** 39–47.

Lasley RD, Anderson GM and Mentzer RM Jr. (1993) Ischemic and hypoxic preconditioning enhance postischemic recovery of function in the rat heart. *Cardiovasc. Res.* **27,** 565–570.

Law WR and Raymond RM. (1988) Adenosine potentiates insulin-stimulated myocardial glucose uptake *in vivo*. *Am. J. Physiol.* **254,** H970–H975.

Lawson CS. (1994) Does ischemic preconditioning occur in the human heart? *Cardiovasc. Res.* **28,** 1461–1466.

Lawson CS, Coltart DJ and Hearse DJ. (1993) The antiarrythmic action of ischemic preconditioning in rat hearts does not involve functional G_i proteins. *Cardiovasc. Res.* **27,** 681–687.

Lee HT, LaFaro RJ and Reed GE. (1995) Pretreatment of human myocardium with adenosine during open heart surgery. *J. Card. Surg.* **10,** 665–676.

Li GC, Vasquez JA, Gallagher KP and Lucchesi BR. (1990) Myocardial protection with preconditioning. *Circulation* **82,** 609–619.

Li R-K, Weisel RD, Williams WG and Mickle DAG. (1992) Method of culturing cardiomyocytes from human pediatric ventricular myocardium. *J. Tiss. Cult. Meth.* **14,** 93–100.

Li R-K, Shaikh N, Weisel RD, Williams WG and Mickle DAG. (1994) Oxyradical-induced antioxidant and lipid changes in cultured human cardiomyocytes. *Am. J. Physiol.* **266,** H2204–H2211.

Li Y and Kloner RA. (1992) Cardioprotective effects of ischaemic preconditioning are not mediated by prostanoids. *Cardiovasc. Res.* **26,** 226–231.

Limas CJ and Limas C. (1983) Involvement of microtubules in the isoproterenol-induced 'down'-regulation of myocardial β-adrenergic receptors. *Biochim. Biophys. Acta* **735,** 181–184.

Liu GS, Stanley AW and Downey J. (1991a) Cyclooxygenase products are not involved in the protection against myocardial infarction afforded by preconditioning in the rabbit. Cyclooxygenase pathway's involvement in preconditioning. *Am. J. Cardiovasc. Pathol.* **4,** 157–164.

Liu GS, Thornton J, Van Winkle DM, Stanley AWH, Olsson RA and Downey JM. (1991b) Protection against infarction afforded by preconditioning is mediated by A_1 adenosine receptors in rabbit heart. *Circulation* **84,** 350–356.

Liu GS, Stanley AW and Downey JM. (1992) Ischaemic preconditioning is not dependent on neutrophils or glycolytic substrate at reperfusion in rabbit heart. *Cardiovasc. Res.* **26,** 1195–1198.

Liu GS, Richards SC, Olsson RA, Mullane K, Walsh RS and Downey JM. (1994) Evidence that the adenosine A_3 receptor may mediate the protection afforded by preconditioning in the isolated rabbit heart. *Cardiovasc. Res.* **28,** 1057–1061.

Liu Y and Downey JM. (1992) Preconditioning protects against infarction in the rat heart. *Am. J. Physiol.* **263,** H1107–H1112.

Liu Y and Downey JM. (1993) Preconditioning against infarction in the rat heart does not involve a pertussis toxin sensitive G protein. *Cardiovasc. Res.* **27,** 608–611.

Liu Y, Ytrehus K and Downey JM. (1994) Evidence that translocation of protein kinase C is a key event during ischemic preconditioning of rabbit myocardium. *J. Mol. Cell. Cardiol.* **26,** 661–668.

Lu H, Remeysen P and De Clerck F. (1993) The protection by ischemic preconditioning against myocardial ischemia- and reperfusion-induced arrhythmias is not mediated by ATP-sensitive potassium channels in rats. *Cor. Art. Dis.* **4,** 649–657.

Mainwaring R, Lasley R, Rubio R, Wyatt DA and Mentzer RM Jr. (1988) Adenosine stimulates glucose uptake in the isolated rat heart. *Surgery* **13,** 445–449.

Maggirwar SB, Dhanraj DN, Somani SM and Ramkumar V. (1994) Adenosine acts as an endogenous mediator of the cellular antioxidant defense system. *Biochem. Biophys. Res. Commun.* **201,** 508–515.

Marber MS, Latchman DS, Walker JM and Yellon DM. (1993) Cardiac stress protein elevation 24 hours after brief ischemia or heat stress is associated with resistance to myocardial infarction. *Circulation* **88,** 1264–1272.

Michell RH. (1975) Inositol phospholipids and cell surface receptor function. *Biochim. Biophys. Acta* **415,** 81–87.

Mickle DAG, Li R-K, Weisel RD, Tumiati LC and Wu FW. (1990) Water soluble antioxidant specificity against free radical injury using cultured human ventricular myocytes and fibroblasts and saphenous vein endothelial cells. *J. Mol. Cell. Cardiol.* **22,** 1297–1304.

Mitchell MB, Winter CB, Locke-Winter CR, Banerjee A and Harken AH. (1993) Cardiac preconditioning does not require myocardial stunning. *Ann. Thorac. Surg.* **55,** 395–400.

Mitchell MB, Meng X, Ao L, Brown JM, Harken AH and Banerjee A. (1995) Preconditioning of isolated rat heart is mediated by protein kinase C. *Circ. Res.* **76,** 73–81.

Mochly-Rosen D, Khaner H, Lopez J and Smith BL. (1992) Intracellular receptor for activated protein kinase C. Identification of binding site for the enzyme. *J. Biol. Chem.* **266,** 14 866–14 868.

Moses RL and Claycomb WC. (1989) Differential membrane specializations and myofibrillar breakdown and recovery in cultured adult cardiac myocytes treated with PMA and diacyglycerol. *J. Cell. Sci.* **93,** 93–105.

Munishi R and Linden J. (1989) Co-purification of A_1 adenosine receptors and guanine nucleotide-binding proteins from bovine brain. *J. Biol. Chem.* **264,** 14 853–14 859.

Murphy E, Fralix TA, London RE and Steenbergen C. (1993) Effects of adenosine antagonists on hexose uptake and preconditioning in perfused rat heart. *Am. J. Physiol.* **265,** C1146–C1155.

Murry CE, Jennings RB and Reimer KA. (1986) Preconditioning with ischemia: a delay of lethal cell injury in ischemic myocardium. *Circulation* **74,** 1124–1136.

Murry CE, Richard VJ, Reimer KA and Jennings RB. (1990) Ischemic preconditioning slows energy metabolism and delays ultrastructural damage during a sustained ischaemic episode. *Circ. Res.* **66,** 913–931.

Nakhostine N and Lamontagne D. (1993) Adenosine contributes to hypoxia-induced vasodilatation through ATP-sensitive K^+ channel activation. *Am. J. Physiol.* **265,** H1289–H1293.

Neely CF, Haile DM, Cahill BE and Kadowitz PJ. (1991) Adenosine and ATP produce vasoconstriction in the feline pulmonary vascular bed by different mechanisms. *J. Pharmacol. Exp. Ther.* **258,** 753–761.

Newby AC. (1984) Adenosine and the concept of 'retaliatory metabolites'. *Trends Biochem. Sci.* **9,** 42–44.

Nishizuka Y. (1986) Studies and perspectives of protein kinase C. *Science* **233,** 305–312.

Nishizuka Y. (1988) The molecular heterogeneity of protein kinase C and implications for cellular regulation. *Nature* **334,** 661–665.

Nishizuka Y. (1992) Intracellular signalling by hydrolysis of phospholipids and activation of protein kinase C. *Science* **258,** 607–614.

Okajima F, Sato K, Sho K and Kondo Y. (1989) Stimulation of adenosine receptor enhances α_1-adrenergic receptor-mediated activation of phospholipase C and Ca^{2+} mobilization in pertussis toxin-sensitive manner in FRTL-5 thyroid cells. *FEBS Lett.* **248,** 145–149.

Olsson RA and Pearson JD. (1990) Cardiovascular purinoceptors. *Physiol. Rev.* **70,** 761–845.

Omar BA, Hanson AK, Bose SK and McCord JM. (1991) Ischemic preconditioning is not mediated by free radicals in the isolated rabbit heart. *Free Rad. Biol. Med.* **11,** 517–520.

Osada M, Takeda S, Sato T, Komori S and Tamura K. (1994) The protective effect of preconditioning on reperfusion-induced arrhythmia is lost by treatment with superoxide dismutase. *Jpn. Circ. J.* **58,** 259–263.

Piacentini L, Wainwright CL and Parratt JR. (1993) The antiarrythmic effect of ischaemic preconditioning in isolated rat heart involves a pertussis toxin sensitive mechanism. *Cardiovasc. Res.* **27,** 689–690.

Pollard JK and Mitchell MD. (1993) Tumor necrosis factor alpha stimulates amnion prostaglandin biosynthesis primarily via an action on fatty acid cyclooxygenase. *Prostaglandins* **46,** 499–510.

Przyklenk K, Bauer B, Ovize M, Kloner RA and Whittaker P. (1993) Regional ischemic 'preconditioning' protects remote virgin myocardium from subsequent sustained coronary occlusion. *Circulation* **87,** 893–899.

Rasmussen H, Calle RA, Ganesan S, Smallwood JI, Throckmorton DC and Zawalich WS. (1992) Protein kinase C: role in sustained cellular responses. In: *Protein kinase C. Current Concepts and Future Perspectives* (eds DS Lester and RM Epand). Ellis Horwood, London, pp. 183–236.

Richard V, Tron C and Thuillez C. (1993) Ischaemic preconditioning is not mediated by oxygen derived free radicals in rats. *Cardiovasc. Res.* **27,** 2016–2021.

Sandhu R, Diaz RJ and Wilson GJ. (1993) Comparison of ischaemic preconditioning in blood and buffer perfused heart models. *Cardiovasc. Res.* **27,** 602–607.

Schiemann W, Westfall D and Buxton L. (1991) Smooth muscle adenosine A_1 receptors couple to disparate effectors by distinct G proteins in pregnant myometrium. *Am. J. Physiol.* **261,** E141–E150.

Schott RJ, Rohmann S, Braun ER and Schaper W. (1990) Ischemic preconditioning reduces infarct size in swine myocardium. *Circ. Res.* **66,** 1133–1142.

Schwibert E, Karlson H, Friedman P, Dietl P and Speilman W. (1992) Adenosine regulates a chloride channel via protein kinase C and a G protein in a rabbit cortical collecting duct cell line. *J. Clin. Invest.* **89,** 834–841.

Shiki K and Hearse DJ. (1987) Preconditioning of ischemic myocardium: reperfusion-induced arrhythmias. *Am. J. Physiol.* **253,** H1470–H1476.

Speechly-Dick ME, Mocanu MM and Yellon DM. (1994) Protein kinase C. Its role in ischemic preconditioning in the rat. *Circ. Res.* **75,** 586–590.

Speechly-Dick ME, Grover GJ and YUellon DM. (1995) Does ischemic preconditioning in the human involve protein kinase C and the ATP-dependent K^+ channel? Studies of contractile function following simulated ischemia in an atrial *in vitro* model. *Circ. Res.* **77,** 1030–1035.

Spielman W, Klotz K, Arend L, Olson B, LeVier D and Schwabe U. (1992) Characterization of adenosine A1 receptor in a cell line (28A) derived from a rabbit collecting tubule. *Am. J. Physiol.* **263,** C502–C508.

Steenbergen C, Fralix TA and Murphy E. (1993a) Role of increased cytosolic free calcium concentration in myocardial ischemic injury. *Basic Res. Cardiol.* **88,** 456–470.

Steenbergen C, Perlman ME, London RE and Murphy E. (1993b) Mechanism of preconditioning: ionic alterations. *Cardiovasc. Res.* **72,** 112–125.

Stokke M, Aksnes G, Lande K, Hagelin EM and Brors O. (1994) Density of L-type calcium channels in ischaemically preconditioned porcine heart regions. *Acta Phys. Scand.* **150,** 425–430.

Streb H, Irvine RF, Berridge MJ and Schulz I. (1983) Release of Ca^{2+} from a non-mitochondrial intracellular store in pancreatic acinar cells by inositol 1,4,5-trisphosphate. *Nature* **306,** 67–69.

Sutherland EW. (1972) Studies on the mechanism of hormone action. *Science* **177,** 401–408.

Takai Y, Kishimoto A, Inoue M and Nishizuka Y. (1977) Studies on a cyclic nucleotide-independent protein kinase and its proenzyme in mammalian tissues. I. Purification and characterization of an active enzyme from bovine cerebellum. *J. Biol. Chem.* **252,** 7603–7609.

Takai Y, Kishimoto A, Kikkawa U, Mori T and Nishizuka Y. (1979) Unsaturated diacylglycerol as a possible messenger for the activation of calcium-activated, phospholipid-dependent protein kinase system. *Biochem. Biophys. Res. Commun.* **91,** 1218–1224.

Tan HL, Mazon P, Verberne HJ, Sleeswijk ME, Coronel R, Opthof T and Janse MJ. (1993) Ischaemic preconditioning delays ischaemia induced cellular electrical uncoupling in rabbit myocardium by activation of ATP sensitive potassium channels. *Cardiovasc. Res.* **27,** 644–651.

Thornton J, Striplin S, Liu GS, Swafford A, Stanley AWH, Van Winkle DM and Downey JM. (1990) Inhibition of protein synthesis does not block myocardial protection by preconditioning. *Am. J. Physiol.* **259,** H1822–H1825.

Thornton JD, Liu GS, Olsson RA and Downey JM. (1992) Intravenous pretreatment with A_1-selective adenosine analogues protects the heart against infarction. *Circulation* **85,** 659–665.

Thornton JD, Liu GS and Downey JM. (1993a) Pretreatment with pertussis toxin blocks the protective effects of preconditioning: evidence for a G-protein mechanism. *J. Mol. Cell. Cardiol.* **25,** 311–320.

Thornton JD, Thornton CS, Sterling DL and Downey JM. (1993b) Blockade of ATP-sensitive potassium channels increases infarct size but does not prevent preconditioning in rabbit hearts. *Circ. Res.* **72,** 44–49.

Tominaga K, Shibata S and Watanabe S. (1992) A neuroprotective effect of adenosine A1-receptor agonists on ischemia-induced decrease in 2-deoxyglucose uptake in rat hippocampal slices. *Neurosci. Lett.* **145,** 67–70.

Toombs CF, Moore TL and Shebuski RJ. (1993) Limitation of infarct size in the rabbit by ischaemic preconditioning is reversible with glibenclamide. *Cardiovasc. Res.* **27,** 617–622.

Tosaki A, Cordis GA, Szerdahelyi P, Engelman RM and Das DK. (1994) Effects of preconditioning on reperfusion arrhythmias, myocardial functions, formation of free radicals, and ion shifts in isolated ischemic/reperfused rat hearts. *J. Cardiovasc. Pharmacol.* **23**, 365–373.

Tsuchida A, Thompson R, Olsson RA and Downey JM. (1994) The anti-infarct effect of an adenosine A_1-selective agonist is diminished after prolonged infusion as is the cardioprotective effect of ischaemic preconditioning in rabbit heart. *J. Mol. Cell. Cardiol.* **26**, 303–311.

Tumiati LC, Li R-K, Weisel RD, Williams WG and Mickle DAG. (1994) An *in vitro* model to study myocardial ischemic injury. *J. Tiss. Cult. Meth.* **16**, 1–9.

Tuurens JF, Thornton J, Barnard ML, Snyder S, Liu G and Downey JM. (1992) Protection from reperfusion injury by preconditioning does not require increased antioxidant defences. *Am. J. Physiol.* **262**, H585–H589.

Urabe K, Miura T, Iwamoto T, Ogawa T, Goto M, Sakamoto J and Iimura O. (1993) Preconditioning enhances myocardial resistance to postischaemic myocardial stunning via adenosine receptor activation. *Cardiovasc. Res.* **27**, 657–662.

Vander Heide RS, Reimer KA and Jennings RB. (1993) Adenosine slows ischaemic metabolism in canine myocardium *in vitro*; relationship to ischaemic preconditioning. *Cardiovasc. Res.* **27**, 669–673.

Van Winkle DM, Chien GL, Wolff RA, Soifer BE, Kuzume K and Davis RF. (1994) Cardioprotection provided by adenosine receptor activation is abolished by blockade of the K_{ATP} channel. *Am. J. Physiol.* **266**, H829–H839.

Van Wylen DG. (1994) Effect of ischemic preconditioning on interstitial purine metabolite and lactate accumulation during myocardial ischemia. *Circulation* **89**, 2283–2289.

Vegh A, Papp JG, Szekeres L and Parratt JR. (1990a) Are ATP sensitive potassium channels involved in the pronounced antiarrhythmic effects of preconditioning? *Cardiovasc. Res.* **27**, 638–643.

Vegh A, Szekeres L and Parratt JR. (1990b) Protective effects of preconditioning involve cyclo-oxygenase products. *Cardiovasc. Res.* **24**, 1020–1023.

Vergauwen L, Hespel P and Richter EA. (1994) Adenosine receptors mediate synergistic stimulation of glucose uptake and transport by insulin and by contractions in rat skeletal muscle. *J. Clin. Invest.* **93**, 974–981.

Walker DM, Walker JM, Pattison C, Pugslet W and Yellon DM. (1993) Preconditioning protects isolated human muscle. *J. Mol. Cell. Cardiol.* **27**, 1349–1357.

Williams DO, Boatwright RB, Rugh KS, Gatner HE and Griggs DM Jr. (1991) Myocardial blood flow, metabolism, and function with repeated coronary occlusion in conscious ponies. *Am. J. Physiol.* **260**, H100–H109.

Williamson JR, Cooper RH, Joseph SK and Thomas AP. (1985) Inositol trisphosphate and diacylglycerol as intracellular second messengers in liver. *Am. J. Physiol.* **248**, C203–C216.

Wolfe CL, Sievers RE, Visseren FLJ and Donnelly TJ. (1993) Loss of myocardial protection after preconditioning correlates with the time course of glycogen recovery within the preconditioned segment. *Circulation* **87**, 881–892.

Wyatt DA, Edmunds MC, Rubio R and Berne RM. (1989) Adenosine stimulates glycolytic flux in isolated perfused rat hearts by A_1-adenosine receptors. *Am. J. Physiol.* **257**, H1952–H1957.

Yau TM, Mickle DAG and Weisel RD. (1990) Myocardial free radical reperfusion injury during cardiac surgery. In: *Frontiers in Cardiovascular Research: Cardiac Surgery, State of the Art Reviews* (eds IL Kron and C Mavroudis). Hanley and Belfus Inc., Philadelphia, pp. 703–712.

Yellon DM, Alkhulaifi AM and Pugsley WB. (1993) Preconditioning the human myocardium. *Lancet* **342**, 276–287.

Ytrehus K, Liu Y and Downey JM. (1994) Preconditioning protects ischemic rabbit heart by protein kinase C activation. *Am. J. Physiol.* **266**, H1145–H1152.

Yuan S, Sunahara FA and Sen AK. (1987) Tumor-promoting phorbol esters inhibit cardiac functions and induce redistribution of protein kinase C in perfused beating rat heart. *Circ. Res.* **61**, 372–378.

8

Ischaemic preconditioning during coronary angioplasty

Fabrizio Tomai
Servizio Speciale di Diagnosi e Cura di Emodinamica, Divisione di Cardiochirurgia, Università di Roma Tor Vergata, European Hospital, via Portuense 700, 00149, Rome, Italy

1. Introduction

Ischaemic preconditioning represents the most powerful form of protection against experimental myocardial infarction described to date (Murry *et al.*, 1986). Indeed, this endogenous form of myocardial protection has been shown in all animal species investigated (Cohen *et al.*, 1991; Li *et al.*, 1992; Murry *et al.*, 1986; Schott *et al.*, 1990) and, recently, in isolated human myocytes (Ikonomidis *et al.*, 1994a) and atrial trabeculae (Speechly-Dick *et al.*, 1995; Walker *et al.*, 1995). It seems, therefore, reasonable to assume that such a form of endogenous cardioprotection might also occur in the clinical setting. If this were the case, the possibility of exploiting this endogenous form of protection by pharmacological means would be of great importance in the attempt to reduce myocardial infarct size. However, the traditional experimental model used to study preconditioning, of brief controlled periods of ischaemia to reduce infarction caused by subsequent prolonged ischaemia, cannot be utilized in humans. Thus, ischaemic

Ischaemia: preconditioning and adaptation, edited by M.S. Marber and D.M. Yellon

preconditioning in man has been studied in the following clinical settings: (i) coronary angioplasty, (ii) coronary artery bypass surgery, (iii) exercise-induced ischaemia (warm-up phenomenon), and (iv) pre-infarction angina. The purpose of the present chapter is to review the evidence and mechanisms of adaptation to ischaemia in humans during coronary angioplasty.

2. Coronary angioplasty as a human model for the study of adaptation to ischaemia

Since its introduction in 1976, coronary angioplasty has provided a useful model for studying the effects of transmural myocardial ischaemia due to controlled coronary occlusions in patients with coronary artery disease. In fact, studies during coronary angioplasty have greatly contributed to the understanding of several pathophysiological aspects of myocardial ischaemia in humans, including the role of collateral circulation, stenosis severity and small vessel function (Cohen and Rentrop, 1986; El-Tamini *et al.*, 1991; Rentrop *et al.*, 1985; Tomai *et al.*, 1994a). More recently, the experimental demonstration of preconditioning, together with the common clinical observation of fewer electrocardiographic ischaemic changes and less anginal pain during sequential coronary balloon occlusions, led to the utilization of coronary angioplasty as a model for the study of ischaemic preconditioning in man.

2.1. Angioplasty studies showing adaptation to ischaemia

The first anectodal evidence that an initial coronary balloon inflation had the capacity to improve tolerance to subsequent prolonged coronary occlusion was provided in a case report of a patient who developed an abrupt closure of the left anterior descending coronary artery following the first balloon inflation (Heibig *et al.*, 1989). After 4 min of ischaemia, distal coronary haemoperfusion was started leading to normalization of electrocardiographic changes and disappearance of anginal pain. Interestingly, subsequent balloon inflations of up to 30 min duration did not produce ischaemic changes. However, collateral recruitment, evaluated by means of the distal coronary pressure, increased during very long balloon inflations (from 30 to 42 mmHg during a 10-min balloon inflation). The authors hypothesized that the improved tolerance to ischaemia during prolonged coronary occlusions following a first brief ischaemic period was due to either further recruitment of collateral circulation or to 'conditioning' of the myocardium by the initial ischaemic insult.

The first formal study aimed at assessing adaptation to ischaemia during coronary angioplasty was reported by Deutsch *et al.* (1990) and involved 12 patients with an isolated obstructive stenosis in the left anterior descending coronary artery undergoing two sequential 90-sec balloon inflations. In comparison with the initial balloon occlusion, the second occlusion was characterized by less subjective anginal pain, less ST-segment shift and lower mean pulmonary artery pressure, despite a reduction in cardiac vein flow and unchanged coronary wedge pressure. Furthermore, in an additional group of seven patients, there was less myocardial lactate production during the second inflation (see *Figure 1*).

In agreement with the results of Deutsch *et al.* (1990), we confirmed in a larger patient population involving 48 patients with a single-vessel disease undergoing two sequential 2-min balloon inflations, that the mean ST-segment shift during the second inflation was significantly less than that during the first inflation on both the surface and the intracoronary electrocardiograms (Tomai *et al.*, 1993). Anginal pain was also less severe during the second inflation in patients whose balloon inflations were of equal pressure. It is worth noting that when the pressure used during the second balloon inflation was higher than that used during the first inflation, the severity of the cardiac pain increased, even though the severity of myocardial ischaemia decreased, thus suggesting that the stretching of the coronary artery wall contributes to the extent of cardiac pain experienced by patients during coronary angioplasty. In this study, we did not measure coronary blood flow or distal coronary pressure during balloon inflation. Therefore, we could not rule out the possibility that the decrease in myocardial ischaemia during the second inflation was due to an increase in collateral blood flow rather than to a metabolic myocardial adaptation to ischaemia.

In nine patients undergoing coronary angioplasty for an isolated left anterior descending coronary artery stenosis, Jepson *et al.* (1994) evaluated the release of hypoxanthine [a catabolite of adenosine triphosphate (ATP) and a sensitive marker of ischaemia] immediately after three successive balloon inflations of 60–90, 90–180 and 180–300 sec duration, respectively. Interestingly, hypoxanthine release was reduced after the second and third balloon inflation compared to that after the first inflation, thus confirming an adaptive response of the myocardium to ischaemia.

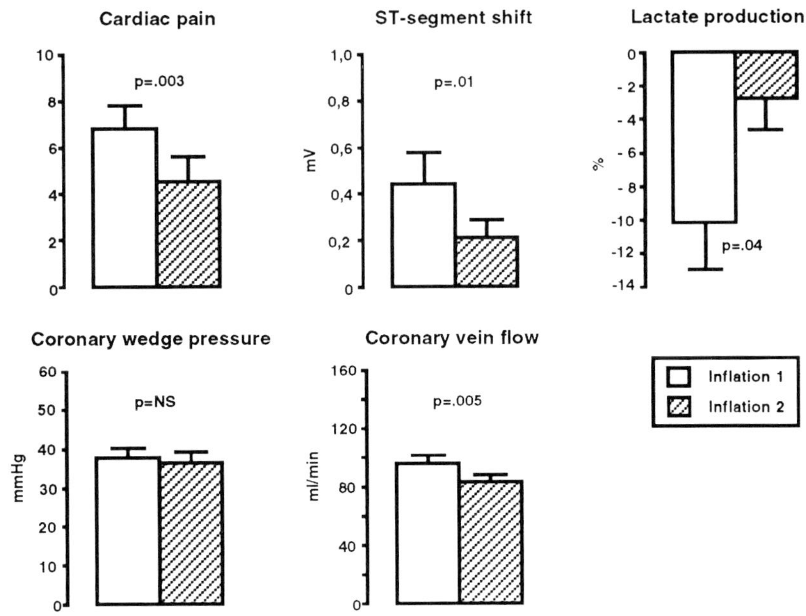

Figure 1. The histograms show the mean values (± 1 SEM) of cardiac pain, ST-segment shift, lactate production, coronary wedge pressure and coronary vein flow at the end of the first (open bars) and second (hatched bars) balloon inflations in the study population. Compared to the first balloon occlusion, the second was characterized by less cardiac pain, ST-segment shift, lactate production and coronary vein flow, while coronary wedge pressure was similar. Adapted from Deutsch *et al.* (1990) with permission from the American Heart Association.

Cribier *et al.* (1994) studied 19 patients with single left anterior descending coronary artery stenosis undergoing four 210 ± 113-sec balloon inflations. When the fourth inflation was compared to the first, they found that onset of cardiac pain was delayed, ST-segment elevation diminished and transmyocardial lactate production decreased. Thus, also in this study a progressive adaptation to repetitive episodes of acute ischaemia could be detected. Daugherty *et al.* (1994) further confirmed a reduction of ischaemic change during repeated balloon inflations, which was independent of the duration of the first balloon inflation. Vaitkus *et al.* (1994) also found a preconditioning-like effect during coronary angioplasty, using a bipolar left ventricular endocardial electrocardiogram.

Some studies, however, failed to show an adaptation to ischaemia following repeated balloon inflations. Oldroyd *et al.* (1992) found no significant difference in the degree of ST-segment shift or in myocardial lactate production between any of three serial balloon inflations, each of 60 sec duration, in 15 patients with stable angina pectoris undergoing angioplasty of a proximal left anterior descending coronary artery stenosis. Similarly, in 28 patients de Jong *et al.* (1993) failed to find significant differences in myocardial lactate and hypoxanthine release among four balloon inflations lasting 60–90 sec. De Bruyne *et al.* (1992) showed less angina following four 60-sec balloon coronary occlusions, but no difference in collateral recruitment (coronary wedge pressure), or in left ventricular function. Dupouy *et al.* (1994), who evaluated left ventricular function after 90 sec of each of three balloon inflations, obtained similar results. Kober *et al.* (1992) also failed to find any significant difference in ST-segment changes between two inflations of 60 sec duration. Finally, Airaksinen *et al.* (1995) found an attenuation of sympathetic activation during the second of two sequential coronary balloon inflations lasting 60–90 sec, but they did not observe any significant change in the severity of myocardial ischaemia. Of note, in all these studies, balloon inflation time was less than or equal to 90 sec.

2.2. Critique of the angioplasty model

Some of the above mentioned studies failed to show adaptation to ischaemia during coronary angioplasty, probably because they neglected some crucial methodological aspects.

Duration of ischaemia. First of all, a critical duration of ischaemia has to be achieved during the first balloon inflation. In agreement with the results in experimental models (Jennings *et al.*, 1992; Lawson and Downey, 1993), a single coronary occlusion longer than 90 sec is required to predictably induce myocardial protection against sequential ischaemic episodes.

Measurement of the severity of myocardial ischaemia. A common criticism of angioplasty studies involving adaptation to ischaemia is the selection of soft endpoints, that is, electrocardiographic changes and anginal pain severity. During coronary angioplasty, the adaptation to ischaemia can be measured by using: (i) anginal pain severity; (ii) surface and intracoronary electrocardiograms; (iii) left ventricular function, and (iv) metabolic changes. The measurement of cardiac pain severity, by using a visual analogue scale, represents a reasonable method for the assessment of directional changes in the severity of ischaemia in individual patients

(Huskisson, 1974). Of note, as during coronary angioplasty, the cardiac pain experienced by patients appears to be also caused by stretching of the coronary arterial wall (Tomai *et al.*, 1993), balloon inflations have to be performed using the same pressure. Thus, because of the subjective nature of cardiac pain, in the absence of additional objective parameters of myocardial ischaemia, this approach does not allow one to draw final conclusions. To this end, it is necessary to show that directional changes of the anginal pain are paralleled by similar changes on the electrocardiogram. Indeed, the measurement of ST-segment shifts on the surface 12-lead and the intracoronary electrocardiograms represents a highly sensitive, well accepted, and simple method for the evaluation of myocardial ischaemia severity during coronary angioplasty (Feldman *et al.*, 1986; Friedman *et al.*, 1986; Naslund *et al.*, 1993; Wohlgelernter *et al.*, 1986). In particular, Labovitz *et al.* (1987) found that during coronary angioplasty ST-segment changes always precede left ventricular dysfunction. Moreover, ST-segment changes are compared in the same patient, where with other variables constant, the most important parameter determining the degree of ST-segment shift appears to be the severity of myocardial ischaemia (Angell *et al.*, 1975; Khuri *et al.*, 1975). Very recently, Lawson *et al.* (1995) and Shattock *et al.* (1995) have measured ST-segment changes in open-chest pigs, subjected to two cycles of 8-min ischaemia, induced by occlusion of left anterior descending coronary artery, and 8-min reperfusion followed by 60-min ischaemia and 2-h reperfusion. They found that ST-segment changes during the first 3 min of ischaemia were less during the second and third ischaemic cycles than during the first, in the absence of a significant increase in collateral flow; they concluded that ST-segment changes provide a reliable index of preconditioning during the first few minutes of coronary occlusion.

The assessment of left ventricular function should, in theory, allow a more direct assessment of the severity of myocardial ischaemia during repeated coronary balloon inflations. Indeed, a transient impairment of global and regional left ventricular function is known to occur during coronary angioplasty (Labovitz *et al.*, 1987; Serruys *et al.*, 1984; Wijns *et al.*, 1986). However, currently available techniques may not be sensitive enough to detect small changes during sequential balloon inflations. This lack of sensitivity may be responsible for some of the controversial results obtained in the evaluation of left ventricular function during coronary angioplasty. For instance, Deutsch *et al.* (1990) failed to find significant changes in left ventricular filling pressure between repeated balloon inflations, despite the decreased clinical, electrocardiographic and metabolic evidence of myocardial ischaemia during the second inflation. Similarly, De Bruyne *et al.* (1992) and Dupouy *et al.* (1994) failed to detect significant differences among sequential coronary balloon inflations, using a tip micromanometer positioned in the left ventricle. Furthermore, the balloon inflation time utilized in these studies was too short (≤ 90 sec), so that the severity of myocardial ischaemia needed to induce detectable changes in left ventricular function was probably not reached. In contrast, Cribier *et al.* (1992) found a reduction of left ventricular end-diastolic pressure and an increase of ejection fraction during the fifth balloon inflation, compared to those observed during the second inflation, although they failed to show significant changes of regional wall motion (see *Figure 2*). Thus, it would appear that a convincing demonstration of changes in left ventricular function during sequential balloon inflations requires techniques more sophisticated than those currently utilized (Kass *et al.*, 1990; Serruys *et al.*, 1984).

Finally, changes in the severity of myocardial ischaemia following repeated coronary occlusions can be evaluated by measuring the transcardiac release of lactate, adenosine and/or adenosine catabolites. Indeed, in experimental studies ischaemic preconditioning has been shown to cause preservation of high energy phosphates, notably ATP, and a reduction in the rate of anaerobic glycolysis with reduced lactate accumulation in the preconditioned myocardium (Murry *et al.*, 1990). However, the metabolism of high energy phosphates is difficult to investigate in man (Hill *et al.*, 1992). Conversely, the transcardiac release of lactate, an index of anaerobic glycolysis, is easy to measure but it does not reliably reflect changes in the severity or amount of ischaemia (Apstein *et al.*, 1979; Gertz *et al.*, 1980). Nonetheless, a progressive decrease of transmyocardial lactate production following sequential balloon inflations was actually found in the angioplasty studies in which coronary occlusions longer than 90 sec were employed (Cribier *et al.*, 1994; Deutsch *et al.*, 1990). In conclusion, the best way to assess the severity of myocardial ischaemia during repeated balloon inflations is probably the utilization of multiple endpoints, including the intracoronary electrocardiogram, which remains the most reliable and sensitive tool for the detection of changes in the severity of myocardial ischaemia within the same patient.

Pre-inflation ischaemia. A last, but not least, critical detail in angioplasty studies aimed at assessing adaptation to ischaemia is the achievement of a similar duration of ischaemia during the first and the following inflations, that is, pre-inflation ischaemia has to be avoided. With regard

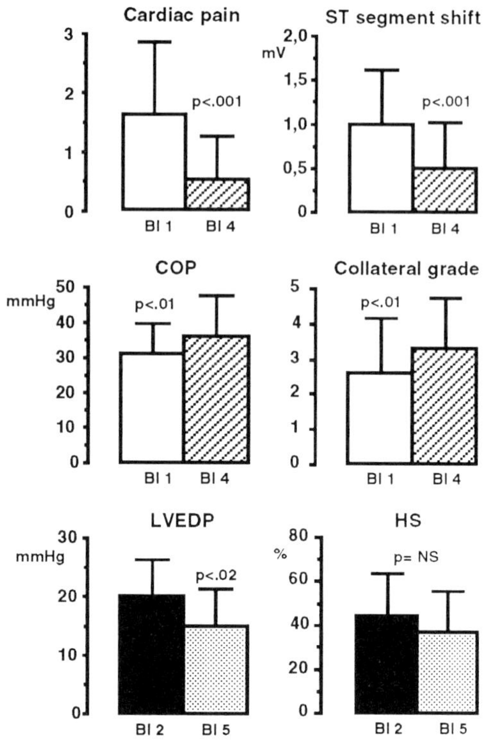

Figure 2. The histograms show the mean values (± 1 SD) of cardiac pain, ST-segment shift, coronary occlusion pressure (COP) and collateral grade at the end of the first (open bars) and fourth (hatched bars) balloon inflations, and of left ventricular end-diastolic pressure (LVEDP) and extent of hypokinetic segments (HS) at the end of the second (solid bars) and fifth (dotted bars) balloon inflations in 17 patients. Compared to the first balloon inflation, the fourth was characterized by less cardiac pain and ST-segment shift, but greater COP and collateral grade. Compared to the second balloon inflation, the fifth was characterized by less LVEDP, but similar HS. BI= balloon inflation. Reprinted with permission from the American College of Cardiology (*Journal of the American College of Cardiology*, 1992, **20**, 578–586).

to this, Denys *et al.* (1992) speculated that the longer time to ischaemia during the second balloon inflation might be due to "pre-inflation ischaemia from decreased blood flow due to obstruction from the uninflated balloon across the lesion rather than to adaptive metabolic myocardial mechanisms". To avoid this confounding factor, we suggest that the guidewire and the balloon catheter should be advanced together, in order to minimize the time needed to put the balloon across the lesion, after that the guidewire is positioned in the distal tract of the stenosed vessel. This expedient is always necessary in patients with tight stenoses, in whom even the guidewire alone could be occlusive. Interestingly, in 10 patients undergoing two sequential inflations, of which the first was 2 min long and the second 3 min long, we found that electrocardiographic ischaemic changes during the second inflation were still less than those during the first (unpublished observations). These results are in agreement with those obtained by Jepson *et al.* (1994), who showed a progressive reduction in the severity of myocardial ischaemia during three sequential balloon inflations, despite a progressive increase in the inflation time.

Residual stenosis at the end of the first inflation. Finally, it is important to achieve a minimal residual stenosis at the end of the first inflation, in order to allow adequate reperfusion and to re-establish baseline haemodynamic and electrocardiographic conditions before the beginning of successive inflations.

3. Mechanisms of adaptation to ischaemia during coronary angioplasty

The adaptation to ischaemia observed following repeated coronary balloon occlusions may be due to either a vascular adaptation, that is, progressive collateral recruitment, or a metabolic myocardial adaptation, that is, ischaemic preconditioning. The latter, in turn, can be mediated by a variety of mechanisms, including: (i) opening of ATP-sensitive potassium (K_{ATP}) channels, (ii) activation of adenosine receptors, (iii) activation of α_1-adrenergic receptors.

3.1. Collateral recruitment during repeated balloon inflations

Coronary collateral flow is a major determinant of the severity of myocardial ischaemia during coronary occlusion, which is known to trigger acute collateral recruitment (Cohen and Rentrop, 1986; Rentrop *et al.*, 1985; Tomai *et al.*, 1994a). Cohen and Rentrop (1986), using contrast injections in the contralateral coronary artery, demonstrated that the improvement of collateral vessel filling observed during coronary angioplasty limits the severity of myocardial ischaemia. More recently, Kyriakides *et al.* (1995) documented vasodilation of collateral donor coronary branch during coronary angioplasty, probably related to an increase in collateral filling. Thus, the adaptation to ischaemia observed during repeated coronary balloon occlusions might be due, theoretically, to a progressive increase in collateral flow. Deutsch *et al.* (1990), however, showed adaptation to ischaemia despite a reduction in cardiac vein flow and unchanged coronary wedge pressure during the second balloon inflation, thus concluding that the adaptation to ischaemia was due to ischaemic myocardial preconditioning rather than to progressive collateral recruitment (see *Figure 1*). Their results contrast with those of Cribier *et al.* (1992), who assessed collateral recruitment by using ipsilateral and contralateral injections of contrast medium and coronary wedge pressure in 17 patients with isolated stenosis of the left anterior descending coronary artery undergoing five sequential balloon inflations, each maintained for a minimum of 120 sec (range 120–370 sec). They confirmed adaptation to myocardial ischaemia during repetitive coronary occlusions characterized by a significant reduction of angina, ST-segment shift and left ventricular filling pressure, but they also found an increase of collateral angiographic grade during the first coronary occlusion and a further increase during the fourth coronary inflation in 10 patients. Accordingly, in these 10 patients coronary wedge pressure slightly increased from the first to the fourth coronary occlusion (see *Figure 2*). Of note, about one half of their patients did not exhibit progressive collateral recruitment, despite a reduction in ischaemic changes during repeated balloon inflations. Furthermore, the progressive increase in recruitable collateral vessels between the first and the fourth inflation observed in 10 patients was very small: the increase in coronary wedge pressure, although statistically significant, was on average only 5 mmHg. On the basis of these findings, Cribier *et al.* (1992) suggested the possibility that both ischaemic preconditioning and progressive collateral recruitment might explain the adaptation to ischaemia observed during coronary angioplasty.

Kyriakidis *et al.* (1994) assessed collateral recruitment by using a Doppler flow velocimeter positioned in the proximal right coronary artery and quantitative coronary angiography in 24 patients undergoing four sequential 90-sec balloon inflations for single left anterior descending coronary artery stenoses. During the initial inflation, 14 patients developed angiographically visible recruitable vessels, whereas 10 patients did not. No further recruitment could be detected in any patient at angiography during the successive balloon inflations. The mean right coronary artery blood flow velocity transiently increased to a nearly identical level during all four balloon inflations in six patients who developed angiographically evident high grade recruitable collateral vessels; in eight patients who developed angiographically evident low grade recruitable collateral vessels, it increased only during the third and fourth occlusions. Finally, it did not change from baseline in the remaining 10 patients without angiographically evident recruitable vessels (see *Figure 3*). Thus, only about 30% of the patients actually exhibited progressive collateral recruitment after the first inflation. Furthermore, in Kyriakidis' study (Kyriakidis *et al.*, 1994), similar to that seen by Cribier *et al.* (1992), this progressive increase in collateral flow could be detected only after several balloon inflations, while in most angioplasty studies aimed at assessing ischaemic preconditioning, only the first two or three inflations were taken into account (Claeys *et al.*, 1996; Deutsch *et al.*, 1990; Tomai *et al.*, 1994b, 1996).

We recently studied nine patients undergoing single vessel coronary angioplasty, by using a 0.014-inch Doppler intracoronary guidewire, positioned in collateral donor coronary branch (Tomai *et al.*, 1995). The

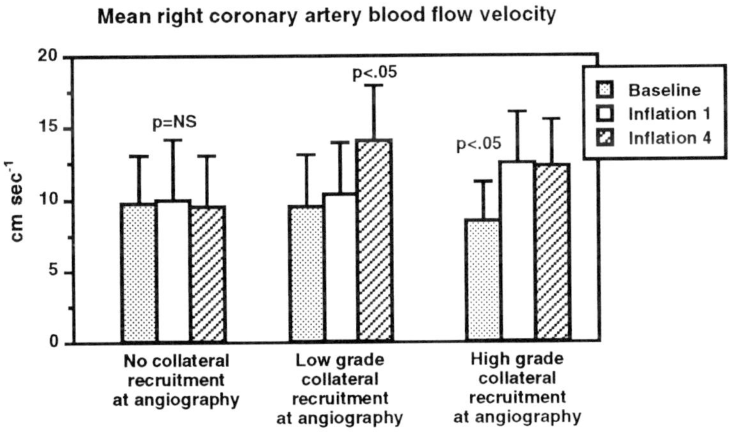

Figure 3. The histograms show the mean values (± 1 SD) of right coronary artery blood flow velocity at baseline (dotted bars) and during the first (open bars) and fourth (hatched bars) balloon inflations in three groups of patients. The mean right coronary artery blood flow velocity during the first and fourth inflations did not change from baseline in 10 patients without angiographically evident recruitable collateral vessels; it progressively increased in eight patients who developed angiographically evident low grade recruitable collateral vessels, with significant increments only during the fourth inflation; it transiently increased to a nearly identical level during both the first and fourth inflations in six patients who developed angiographically evident high grade recruitable collateral vessels. Reprinted with permission of the American College of Cardiology (Kyriakidis *et al.*, *Journal of the American College of Cardiology*, 1994, **23**, 869–878).

mean values of ST-segment shifts on the surface 12-lead and the intracoronary electrocardiograms and the severity of cardiac pain were measured at the end of the first and second balloon inflations, both 2 min long, separated by at least 5 min of reperfusion. The mean ST-segment shift and anginal pain severity during the second inflation were significantly less than those during the first inflation, while coronary blood flow velocity showed a significant increase during the first inflation, but it did not show a further increase during the second inflation (see *Figure 4*).

In summary, collateral recruitment during a first coronary balloon inflation does occur, while further recruitment during following inflations is infrequent and occurs only after multiple inflations. These findings indicate that in man the adaptation to ischaemia observed in the majority of angioplasty studies, during the first two or three inflations, is mainly mediated by ischaemic myocardial preconditioning rather than by progressive collateral recruitment.

3.2. Ischaemic preconditioning

Further evidence that the adaptation to ischaemia during repeated balloon inflations is mainly mediated by ischaemic preconditioning is

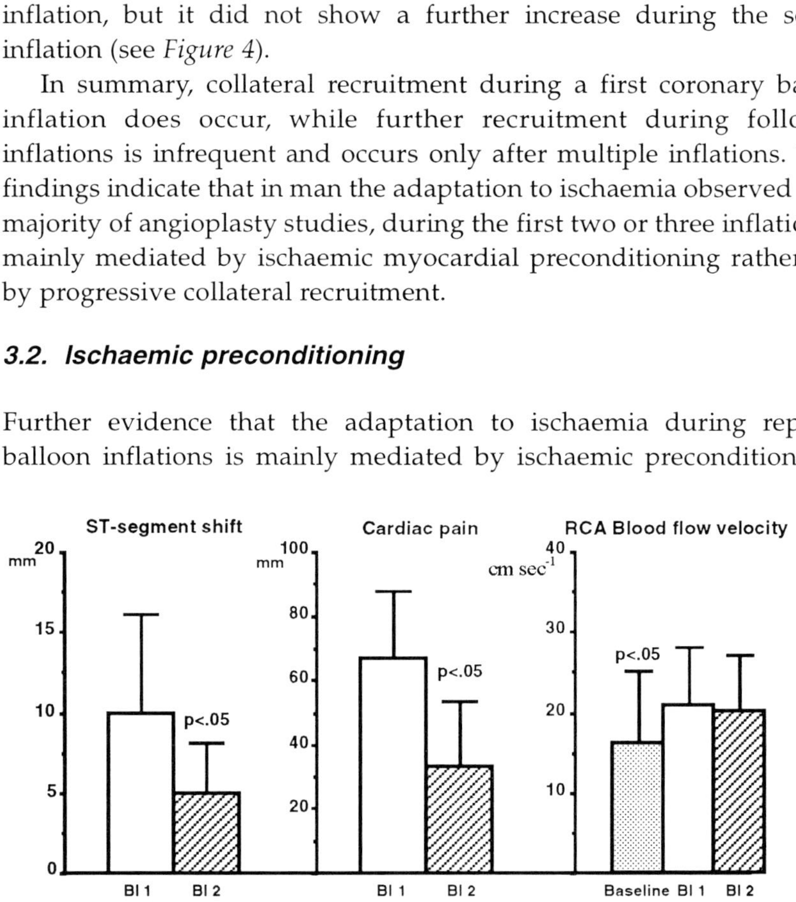

Figure 4. The histograms show the mean values (± 1 SD) of ST-segment shift on the intracoronary electrocardiogram, cardiac pain severity and right coronary artery (RCA) blood flow velocity (average peak velocity) at the end of the first (open bars) and second (hatched bars) balloon inflations in nine patients. Compared to the first balloon inflation, the second was characterized by less ST-segment shift and cardiac pain severity. Of note, baseline (dotted bars) RCA blood flow velocity showed a significant increase during the first inflation, but it did not further increase during the second inflation. BI= balloon inflation.

derived from the fact that preconditioning in the patient can be prevented by agents which specifically prevent ischaemic preconditioning in experimental models (Downey *et al.*, 1994; Gross *et al.*, 1994; Walker and Yellon, 1992), in human myocytes (Ikonomidis *et al.*, 1994b, 1995) or atrial trabeculae (Speechly-Dick *et al.*, 1995; Walker *et al.*, 1995).

K_{ATP} channels. To establish whether the reduction of myocardial ischaemia observed in humans during coronary angioplasty following repeated balloon inflations is due to activation of K_{ATP} channels, we randomized 20 consecutive patients undergoing one-vessel coronary angioplasty to receive 10 mg oral glibenclamide, a selective K_{ATP} channel blocker, or placebo (Tomai *et al.*, 1994b). The mean values of ST-segment shifts on the surface 12-lead and the intracoronary electrocardiograms were measured at the end of the first and second balloon inflations, both 2 min long. We found that in glibenclamide-treated patients, the mean ST-segment shift during the second balloon inflation was similar to that observed during the first inflation, and the severity of cardiac pain was even greater. Conversely, in placebo-treated patients the mean ST-segment shift during the second inflation was less than that during the first inflation, as was the severity of cardiac pain (see *Figure 5*). Therefore, as the ischaemic preconditioning observed during brief repeated coronary occlusions was completely abolished by pre-treatment with glibenclamide, we suggested that in this human model it is predominantly mediated by K_{ATP} channels. As we did not measure collateral recruitment, we could not exclude the theoretical possibility that glibenclamide prevented collateral recruitment. Indeed, high doses of intracoronary glibenclamide are known to decrease coronary blood flow and interfere with hypoxic and ischaemic vasodilation by blockade of K_{ATP} channels in vascular smooth muscle (Daut *et al.*, 1990; Imamura *et al.*, 1992). However, if glibenclamide had prevented ischaemic vasodilation, we should also have obtained greater electrocardiographic changes and more severe pain in glibenclamide-treated patients at the end of the first inflation. Instead, the magnitude of ischaemic electrocardiographic changes and the severity of pain at the end of the first inflation were similar in glibenclamide- and placebo-treated patients.

A role of K_{ATP} channels in ischaemic preconditioning in humans has also been suggested by Taggart *et al.* (1993), who assessed myocardial action potentials, which are known to shorten during myocardial

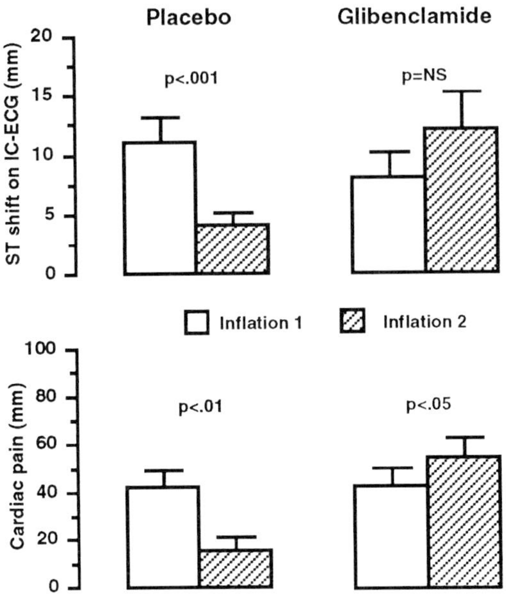

Figure 5. The histograms show the mean values (± 1 SEM) of ST-segment shift on the intracoronary electrocardiogram (IC-ECG) and cardiac pain severity at the end of the first (open bars) and second (hatched bars) balloon inflations in placebo- or glibenclamide-treated patients. In placebo-treated patients, ST-segment changes and cardiac pain severity at the end of the second balloon inflation were significantly less than those at the end of the first inflation. Conversely, in glibenclamide-treated patients, ST-segment changes and cardiac pain severity at the end of the second balloon inflation were similar and, respectively, greater than those at the end of the first inflation. Adapted from Tomai *et al.* (1994b) with permission from the American Heart Association.

ischaemia, in seven patients undergoing coronary angioplasty of the left anterior descending coronary artery with two 90-sec balloon inflations separated by 2 min of reflow. They found a reduction of ST-segment elevation in the intracoronary electrocardiogram during the second inflation, while the action potential duration shortening was similar to that observed during the first inflation. As the activation of the K_{ATP} channel shortens the action potential, a greater activation of these channels during the second inflation may explain why action potential duration failed to increase despite the lesser severity of myocardial ischaemia.

Adenosine receptors. Some recent studies have suggested that adenosine may play an important role in ischaemic preconditioning during coronary angioplasty. In an early report involving 12 patients undergoing coronary angioplasty of the left anterior descending coronary artery with two or three 2-min balloon inflations separated by 3 min of reperfusion, Hashimura *et al.* (1992) showed a reduction in ischaemic electrocardiographic changes, anginal pain and transmyocardial lactate release during the last inflation, compared with the initial inflation. Interestingly, transmyocardial adenosine release was paradoxically increased during the last inflation. The authors concluded that a paradoxical increase of adenosine release from the preconditioned myocardium might account for this phenomenon. Also Inoue *et al.* (1994) found increased adenosine levels in the great cardiac vein after three sequential 1-min balloon inflations, despite the reduction in electrocardiographic ischaemic changes and similar reduction in great cardiac vein oxygen saturations among the inflations. This 'adenosine paradox' also seems to occur in humans during sequential exercise tests (Okazaki *et al.*, 1993) and might be related to an increase in 5'-nucleotidase activity reported to occur in preconditioned myocardium (Kitakaze *et al.*, 1993), rather than reflecting an increased adenosine production. By contrast, experimental studies have clearly shown that in preconditioned myocardium adenosine release is reduced (Mizumura *et al.*, 1995; Van Wylen, 1994; Wikstrom *et al.*, 1995), in agreement with the observation that high energy phosphates are preserved (Murry *et al.*, 1986, 1990). Further studies are warranted to clarify the adenosine paradox.

An important role for adenosine in ischaemic preconditioning during coronary angioplasty is suggested by the results of Claeys *et al.* (1996). They studied 20 patients undergoing three consecutive 90-sec balloon inflations, of which 10 were pre-treated with aminophylline, a non-selective antagonist of adenosine receptors. Although during the third inflation the control patients showed a reduction in the ST-segment shift on the intracoronary electrocardiogram, as compared to the first two inflations, in patients pre-treated with aminophylline, the electrocardiographic ischaemic changes were similar among the inflations (see *Figure 6*, top panel).

In agreement with Claeys *et al.* (1996), we recently demonstrated that bamiphylline, a rather selective A_1 adenosine receptor antagonist (Abbracchio and Cattabeni, 1987), abolished the ischaemic preconditioning observed in man during coronary angioplasty (Tomai *et al.*, 1996). To this end, we studied 36 patients undergoing two sequential 2-min balloon inflations, separated by 5 min of reperfusion. Patients were randomized to receive an intravenous infusion of bamiphylline or

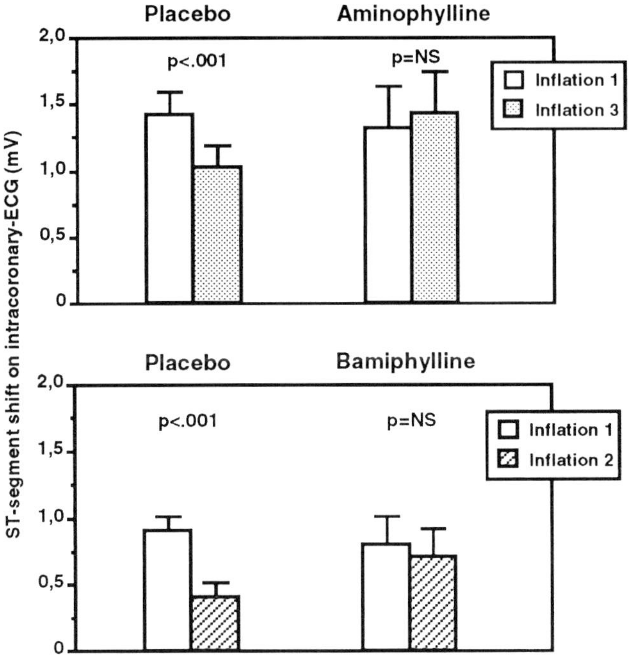

Figure 6. *Top panel*: histograms showing the mean values (± 1 SEM) of ST-segment shift on the intracoronary electrocardiogram (ECG) at the end of the first (open bars) and third (dotted bars) balloon inflations in placebo- or aminophylline-treated patients. In placebo-treated patients, ST-segment changes at the end of the third balloon inflation were significantly less than those at the end of the first inflation, whereas in aminophylline-treated patients there was no significant difference. Adapted from Claeys *et al.* (1996) with permission from Harcourt Brace & Co. Ltd. *Bottom panel*: histograms showing the mean values (± 1 SEM) of ST-segment shift on the intracoronary electrocardiogram (ECG) at the end of the first (open bars) and second (hatched bars) balloon inflations in placebo- or bamiphylline-treated patients. In placebo-treated patients, ST-segment changes at the end of the second balloon inflation were significantly less than those at the end of the first inflation, whereas in bamiphylline-treated patients there was no significant difference. Adapted from Tomai *et al.* (1996) with permission from Harcourt Brace & Co. Ltd.

placebo prior to the procedure. In placebo-treated patients both the mean ST-segment shift on the intracoronary electrocardiogram and the severity of cardiac pain during the second inflation were significantly less than those during the first inflation. Conversely, in bamiphylline-treated patients the mean ST-segment shift and the severity of cardiac pain during the second inflation were similar to those during the first inflation (see *Figure 6*, bottom panel). As bamiphylline is a selective antagonist of A_1 adenosine receptors, these findings suggest that ischaemic preconditioning in man may be mediated by the activation of A_1 adenosine receptors. In this study, we did not assess coronary collateral blood flow. However, we have previously shown that, at the dose used in the present study, bamiphylline does not suppress adenosine-induced cutaneous vasodilation (Pappagallo *et al.*, 1993), nor adenosine-induced coronary vasodilation (Gaspardone *et al.*, 1995a), which are both mediated by vascular A_2 receptors (Belardinelli *et al.*, 1989). Thus, it is very unlikely that in our patients bamiphylline prevented ischaemic preconditioning by limiting an A_2 receptor-mediated increase of coronary blood flow. Finally, we could not exclude the possibility that ischaemic preconditioning was prevented through blockade of A_3 receptors, which appear to mediate preconditioning in rabbits (Armstrong and Ganote, 1994; Liu *et al.*, 1994). In fact, preliminary results suggest that the affinity of bamiphylline for human myocardial A_3 receptors is 27-fold higher than that for A_1 receptors (Mullane, unpublished observations).

α_1-Adrenergic receptors. We are also investigating the role of α_1-adrenergic receptors in ischaemic preconditioning during coronary angioplasty, by using phentolamine mesylate, a selective antagonist of α_1-adrenergic receptors. Early results in the first five patients showed reduction of ST-segment changes and anginal pain severity but unchanged collateral flow following two sequential balloon inflations of 2 min duration, thus suggesting that the activation of α_1-adrenergic receptors does not seem to play an important role in ischaemic preconditioning in this human model (unpublished observations).

4. Conclusions

In man, during coronary angioplasty, both K_{ATP} channels (Taggart *et al.*, 1993; Tomai *et al.*, 1994b) and adenosine receptors, probably A_1 and/or A_3 receptors (Claeys *et al.*, 1996; Tomai *et al.*, 1996), appear to play a key role in mediating ischaemic preconditioning, which seems to be the prevailing cause of the adaptation to ischaemia observed in this human model. How A_1 adenosine receptors and K_{ATP} channels interact in determining preconditioning cannot be deduced from the results of these clinical studies.

The demonstration that glibenclamide prevents ischaemic preconditioning in man (Tomai *et al.*, 1994b) may help to explain the excess mortality from cardiovascular causes observed in diabetic patients on sulphonylureas (University Group Diabetes Program, 1982) and the worse outcome of patients who are on sulphonylureas at the time of acute myocardial infarction (Opie *et al.*, 1979; Rytter *et al.*, 1985). These observations suggest that the treatment of diabetes in some high-risk coronary patients should be shifted from sulphonylureas to insulin, as recently proposed (Hofmann and Opie, 1993). It suggests also that enhancement of preconditioning by K_{ATP} channel openers might limit the detrimental effects of myocardial ischaemia, as previously proposed by both experimental (Gross and Auchampach, 1992; Yao and Gross, 1994) and clinical studies (Saito *et al.*, 1995; Silke *et al.*, 1989; Tamura *et al.*, 1990).

Similarly, the demonstration that adenosine receptor antagonists prevent ischaemic preconditioning (Claeys *et al.*, 1996; Tomai *et al.*, 1996) suggests that methylxanthines should be used with caution in those patients with ischaemic heart disease in whom ischaemic preconditioning is likely to play an important cardioprotective role, that is, those with unstable angina or undergoing coronary artery bypass surgery or coronary angioplasty. On the other hand, adenosine analogues, particularly A_1 (or, perhaps, A_3) receptor agonists, may have the potential to be used as cardioprotective agents, as recently shown in isolated human myocytes (Ikonomidis *et al.*, 1994b, 1995), human atrial trabeculae (Walker *et al.*, 1995), and in experimental (Liu *et al.*, 1991; Thornton *et al.*, 1992) and clinical studies (Lee *et al.*, 1994).

The possibility of fully exploiting preconditioning in humans by pharmacological agents requires a better knowledge of its molecular mechanisms, bearing in mind that the mechanisms of ischaemic preconditioning may differ between different clinical settings. For example A_1/A_3 receptors may be involved in preconditioning during coronary angioplasty (Claeys *et al.*, 1996; Tomai *et al.*, 1996) but not during exercise tests (Gaspardone *et al.*, 1995b; Kerensky *et al.*, 1994).

Acknowledgements

I am grateful to Prof. Pier Agostino Gioffrè and Dr Filippo Crea for constructive criticisms and helpful suggestions.

References

Abbracchio MP and Cattabeni F. (1987) Selective activity of bamiphylline on adenosine A_1-receptors in rat brain. *Pharmacol. Res. Commun.* **19,** 537–545.

Airaksinen KEJ, Ylitalo KV, Penhkurinen KJ, Ikaheimo MJ and Huikuri HV. (1995) Heart rate variability during repeated arterial occlusion in coronary angioplasty. *Am. J. Cardiol.* **75,** 877–881.

Angell CS, Lakatta EG, Weisfeldt ML and Shock NW. (1975) Relationship of intramyocardial oxygen tension and epicardial ST segment changes following acute coronary artery ligation: effect of coronary perfusion pressure. *Cardiovasc. Res.* **9,** 12–18.

Apstein CS, Gravino F and Hood WB. (1979) Limitations of lactate production as an index of myocardial ischemia. *Circulation* **60,** 877–888.

Armstrong S and Ganote CE. (1994) Adenosine receptor specificity in preconditioning of isolated rabbit cardiomyocytes: evidence of A_3 receptor involvement. *Cardiovasc. Res.* **28,** 1049–1056.

Belardinelli L, Linden J and Berne R. (1989) The cardiac effects of adenosine. *Prog. Cardiovasc. Dis.* **32,** 73–97.

Claeys MJ, Vrints CJ, Bosmans JM, Conraads VM and Snoeck JP. (1996) Aminophylline inhibits adaptation to ischemia during angioplasty: role of adenosine in ischemic preconditioning. *Eur. Heart J.* **17,** 538–543.

Cohen M and Rentrop KP. (1986) Limitation of myocardial ischemia by collateral circulation during sudden controlled coronary artery occlusion in human subjects: a prospective study. *Circulation* **74,** 469–476.

Cohen MV, Liu GS and Downey JM. (1991) Preconditioning causes improved wall motion as well as smaller infarcts after transient coronary occlusion in rabbits. *Circulation* **84,** 341–349.

Cribier A, Korsatz L, Koning R, Rath P, Gamra H, Stix G, Merchant S, Chan C and Letac B. (1992) Improved myocardial ischemic response and enhanced collateral circulation with long repetitive coronary occlusion during angioplasty: a prospective study. *J. Am. Coll. Cardiol.* **20,** 578–586.

Cribier A, Eltchaninoff H, Tron C, Koning R, Derumeaux G, Baala B, Farid N, Hecketsweiller B and Letac B. (1994) Adaptation of myocardial ischemia to repetitive prolonged coronary occlusions in PTCA: clinical, ECG, echocardiographic and metabolic assessment. *J. Am. Coll. Cardiol.* (suppl.), 80-A.

Daugherty JAS, Drew BJ, Chou TM, Amidon TM, Ports TA and Wolfe CL. (1994) Electrocardiographic evidence of myocardial preconditioning in patients undergoing PTCA. *J. Am. Coll. Cardiol.* (suppl.), 68-A.

Daut J, Maier-Rudolph W, von Beckerath N, Mehrke G, Gunther K and Goedel-Meinen L. (1990) Hypoxic dilation of coronary arteries is mediated by ATP-sensitive potassium channels. *Science* **247,** 1341–1344.

De Bruyne B, Vantrimpont PJ, Goethals M, Stockbroeckx J and Paulus WJ. (1992) Can coronary angioplasty serve as a model of myocardial preconditioning? *Circulation* **86** (suppl. I), I-378.

de Jong JW, Cargnoni A, Huizer T, Ferrari R and Bonnier JJRM. (1993) Lactate and hypoxanthine release after coronary angioplasty: no evidence of preconditioning. *J. Mol. Cell. Cardiol.* **25** (suppl. I), S-37.

Denys BG, Lee J, Reddy PS and Uretsky BF. (1992) Is there adaptation to ischemia during PTCA? *Eur. Heart J.* **13** (suppl.), 246.

Deutsch E, Berger M, Kussmaul WG, Hirshfeld JW, Herrmann HC and Laskey WK. (1990) Adaptation to ischemia during percutaneous transluminal coronary angioplasty. Clinical, hemodynamic, and metabolic features. *Circulation* **82,** 2044–2051.

Downey JM, Cohen MV, Ytrehus K and Liu. (1994) Cellular mechanisms in ischemic preconditioning: the role of adenosine and protein kinase. *Ann. NY Acad. Sci.* **723,** 882–898.

Dupouy P, Geschwind HJ, Pelle G and Dubois-Rande JL. (1994) Brief repeat human coronary occlusions do not induce myocardial preconditioning. *J. Am. Coll. Cardiol.* (suppl.), 331-A.

El-Tamini H, Davies GT, Sritara P, Hackett D, Crea F and Maseri A. (1991) Inappropriate constriction of small coronary vessels as a possible cause of a positive exercise test early after successful coronary angioplasty. *Circulation* **84,** 2307–2312.

Feldman T, Chua KG and Childers RW. (1986) R wave of the surface and intracoronary electrogram during acute coronary artery occlusion. *Am. J. Cardiol.* **58,** 885–890.

Friedman PL, Shook TL, Kirschenbaum JM, Selwyn AP and Ganz P. (1986) Value of the intracoronary electrocardiogram to monitor myocardial ischemia during percutaneous transluminal coronary angioplasty. *Circulation* **74,** 330–339.

Gaspardone A, Crea F, Tomai F, Versaci F, Iamele M, Gioffrè G, Chiariello L and Gioffrè PA. (1995a) Muscular and cardiac adenosine-induced pain is mediated by A_1 receptors. *J. Am. Coll. Cardiol.* **25,** 251–257.

Gaspardone A, Tomai F and Crea F. (1995b) Clinical studies on adenosine and preconditioning. *J. Mol. Cell. Cardiol.* **27** (suppl. 6), A-351.

Gertz EW, Wisneski JA, Neese R, Houser A, Korte R and Bristow JD. (1980) Myocardial lactate extraction: multi-determined metabolic function. *Circulation* **61,** 256–261.

Gross GJ and Auchampach JA. (1992) Blockade of ATP-sensitive potassium channels prevents myocardial preconditioning in dogs. *Circ. Res.* **70,** 223–233.

Gross GJ, Yao Z, Pieper GM and Auchampach JA. (1994) The ATP-regulated potassium channel in ischemia-reperfusion injury. *Ann. NY Acad. Sci.* **723,** 71–81.

Hashimura K, Matsu'ura Y and Mori I. (1992) Paradoxical release of adenosine in ischemic preconditioning during coronary angioplasty. *Eur. Heart J.* **13** (suppl.), 245.

Heibig J, Bolli R and Harris S. (1989) Initial coronary occlusion improves tolerance to subsequent prolonged balloon inflations. *Cathet. Cardiovasc. Diagn.* **16,** 99–102.

Hill JA, Utterback DB, Keim SG, Dupper D, Mayfield WK, Shryock JC and Belardinelli L. (1992) A catheter–syringe sampling system for measurement of adenosine in coronary sinus blood in humans. *Coronary Artery Dis.* **3,** 963–971.

Hofmann D and Opie LH. (1993) Potassium channel blockade and acute myocardial infarction: implications for management of the non-insulin requiring diabetic patients. *Eur. Heart J.* **14,** 1585–1589.

Huskisson EC. (1974) Measurement of pain. *Lancet* **2,** 1127–1131.

Ikonomidis JS, Tumiati LC, Weisel RD, Mickle DAG and Li RK. (1994a) Preconditioning human ventricular cardiomyocytes with brief periods of simulated ischaemia. *Cardiovasc. Res.* **28,** 1285–1291.

Ikonomidis JS, Shirai T, Weisel RD and Mickle DAG. (1994b) Human cardiomyocyte preconditioning is adenosine A1 receptor dependent. *Circulation* **90** (suppl. I), I-477.

Ikonomidis JS, Shirai J, Weisel RD, Derylo B, Rao V, Whitesite CI, Mickle DAG and Li R-K. (1995) Ischemic or adenosine preconditioning of human ventricular cardiomyocytes is protein kinase C dependent. *Circulation* **92** (suppl. I), I-12.

Imamura Y, Tomoike H, Narishige T, Takahashi T, Kasuya H and Takeshita A. (1992) Glibenclamide decreases basal coronary blood flow in anesthetized dogs. *Am. J. Physiol.* **263,** H399–H404.

Inoue T, Sakai Y, Morooka S, Hayashi T, Takayanagi K and Takabatake Y. (1994) A mechanism of ischemic preconditioning during percutaneous transluminal coronary angioplasty. *J. Am. Coll. Cardiol.* (suppl.), 377-A.

Jennings RB, Murry CE and Reimer KA. (1992) Preconditioning myocardium with ischemia. In: *Stunning, Hibernation, and Calcium in Myocardial Ischaemia and Reperfusion* (ed. LH Opie). Kluwer Academic Publishers, Boston, pp. 154–165.

Jepson NS, Kaisi KK, Cowell RPW, Paul VE, Ilsey CDI and Seymour A-ML. (1994) Reduced hypoxanthine production in human myocardium during PTCA (abstr.). In: *Intrinsic Adaptive Mechanisms During Ischaemia and Reperfusion.* European Heart House, 6–8 October.

Kass DA, Midei M, Brinker J and Maughan WL. (1990) Influence of coronary occlusion during PTCA on end-systolic and end-diastolic pressure–volume relations in humans. *Circulation* **81,** 447–460.

Kerensky RA, Franco EA, Schlaifer JD, Bertolet BD, Belardinelli L and Hill JA. (1994) The effect of intravenous theophylline administration on the warm-up phenomenon during exercise testing in patients with stable angina. *Circulation* **90** (suppl. I), I-497.

Khuri SF, Flaherty JT, O'Riordan JB, Pitt B, Brawley RK, Donahoo JS and Gott VL. (1975) Changes in intramyocardial ST segment voltage and gas tension with regional myocardial ischaemia in the dog. *Circ. Res.* **37,** 455–463.

Kitakaze M, Hori M, Takashima S, Sato H, Inoue M and Kamada T. (1993) Ischemic preconditioning increases adenosine release and 5′-nucleotidase activity during myocardial ischaemia and reperfusion in dogs. *Circulation* **87,** 208–215.

Kober G, Buck T, Sievert H and Vallbracht C. (1992) Myocardial protection during percutaneous transluminal coronary angioplasty: effects of trimetazidine. *Eur. Heart J.* **13,** 1109–1115.

Kyriakidis MK, Petropoulakis PN, Tentolouris CA, Marakas SA, Antonopoulos AG, Kourouclis CV and Toutouzas PK. (1994) Relation between changes in blood flow of the contralateral coronary artery and the angiographic extent and function of recruitable collateral vessels arising from this artery during balloon coronary occlusion. *J. Am. Coll. Cardiol.* **23,** 869–878.

Kyriakides ZS, Tousoulis D, Iliodromitis EK, Apostolou Th, Michelakakis N and Kremastinos DT. (1995) The reactivity of the contralateral artery at the time of balloon dilation during coronary angioplasty. *Eur. Heart J.* **16,** 794–798.

Labovitz AJ, Lewen MK, Kern M, Vandormael M, Deligonal U and Kennedy HL. (1987) Evaluation of left ventricular systolic and diastolic dysfunction during transient myocardial ischemia produced by angioplasty. *J. Am. Coll. Cardiol.* **10,** 748–755.

Lawson CS and Downey JM. (1993) Preconditioning: state of the art myocardial protection. *Cardiovasc. Res.* **27,** 542–550.

Lawson CS, Shattock MJ, Hearse DJ and Downey JM. (1995) Adaptation of ST-segment and monophasic action potential changes during repeated regional ischaemia: preconditioning or enhanced collateral flow? *Eur. Heart J.* **16** (suppl.), 75.

Lee HT, Sanchala VT, Reed GE, Belkin RN and LaFaro RJ. (1994) Preconditioning of human myocardium with adenosine during cardiac surgery. *Circulation* **90** (suppl. I), I-48.

Li YW, Whittaker P and Kloner RA. (1992) The transient nature of the effect of ischemic preconditioning on myocardial infarct size and ventricular arrhythmias. *Am. Heart J.* **123,** 346–353.

Liu GS, Thornton JD, VanWinkle DM, Stanley AWH, Olsson RA and Downey JM. (1991) Protection against infarction afforded by preconditioning is mediated by A_1 adenosine receptors in the rabbit heart. *Circulation* **84,** 350–356.

Liu GS, Richards SC, Olsson RA, Mullane K, Walsh RS and Downey JM. (1994) Evidence that the adenosine A_3 receptor may mediate the protection afforded by preconditioning in the isolated rabbit heart. *Cardiovasc. Res.* **28,** 1057–1061.

Mizumura T, Nithipatikom K and Gross GJ. (1995) Bimakalim, an ATP-sensitive potassium channel opener, mimics the effects of ischemic preconditioning to reduce infarct size, adenosine release, and neutrophil function in dogs. *Circulation* **92,** 1236–1245.

Murry CE, Jennings RB and Reimer KA. (1986) Preconditioning with ischemia: a delay of lethal cell injury in ischemic myocardium. *Circulation* **74,** 1124–1136.

Murry CE, Richard VT, Reimer KA and Jennings RB. (1990) Ischemic preconditioning slows energy metabolism and delays ultrastructural damage during a sustained ischemic episode. *Circ. Res.* **66,** 913–931.

Naslund U, Haggmark S, Johnasson G and Reiz S. (1993) Quantification of myocardium at risk and detection of reperfusion by dynamic vectocardiographic ST-segment monitoring in a pig occlusion–reperfusion model. *Cardiovasc. Res.* **27,** 2170–2178.

Okazaki Y, Kodama K, Sato H, Kitakaze M, Hirayama A, Mishima M, Hori M and Inoue M. (1993) Attenuation of increased regional myocardial oxygen consumption during exercise as a major cause of warm-up phenomenon. *J. Am. Coll. Cardiol.* **21,** 1597–1604.

Oldroyd KG, Paterson JR, Rumley AG, Eteiba H, Rae AP, Shepherd J, Cobbe SM and Hutton I. (1992) Coronary venous lipid peroxide concentrations after coronary angioplasty: correlation with biochemical and electrocardiographic evidence of myocardial ischaemia. *Br. Heart J.* **68,** 43–47.

Opie LH, Tansey MJ and Kenelly BM. (1979) The heart in diabetes mellitus. *S. Afr. Med.* **56,** 256–261.

Pappagallo M, Gaspardone A, Tomai F, Iamele M, Crea F and Gioffrè PA. (1993) Analgesic effect of bamiphylline, a selective antagonist of A1-adenosine receptors, on pain induced by intradermal injection of adenosine. *Pain* **53,** 199–204.

Rentrop KP, Cohen M, Blanke H and Phillips R. (1985) Changes in collateral filling immediately following controlled coronary artery occlusion by an angioplasty balloon in man. *J. Am. Coll. Cardiol.* **5,** 587–592.

Rytter L, Troelsen S and Beck-Nielsen H. (1985) Prevalence and mortality of acute myocardial infarction in patients with diabetes. *Diabetes Care* **8,** 230–234.

Saito S, Mizumura T, Takayama T *et al.* (1995) Anti-ischemic effects of nicorandil during coronary angioplasty in humans. *Cardiovasc. Drugs Ther.* **9,** 257–-263.

Schott RJ, Rohmann S, Braun ER and Schaper W. (1990) Ischemic preconditioning reduces infarct size in swine myocardium. *Circ. Res.* **66,** 1133–1142.

Serruys PW, Wijns W, van den Brand M, Meij S, Slager C, Schuurbiers JCH, Hugenoltz PG and Brower RW. (1984) Left ventricular performance, regional blood flow, wall motion, and lactate metabolism during transluminal angioplasty. *Circulation* **70,** 25–36.

Shattock MJ, Lawson CS, Hearse DJ and Downey JM. (1995) Monophasic action potential and ST-segment changes are indicative of preconditioning only during early ischemia in the pig heart. *Circulation* **92** (suppl. I), I-390.

Silke B, Verma SP, Ali MS, Goldhammer E and Taylor SH. (1989) Effects of nicorandil on left ventricular hemodynamics and volume at rest and during exercise-induced angina pectoris. *Am. J. Cardiol.* **63,** 49J–55 J.

Speechly-Dick ME, Grover GJ and Yellon DM. (1995) Does ischemic preconditioning in the human involve protein kinase C and the ATP-dependent K^+ channel? Studies of contractile function after simulated ischemia in an atrial *in vitro* model. *Circ. Res.* **77,** 1030–1035.

Taggart PI, Sutton PMI, Oliver RM and Swanton RH. (1993) Ischemic preconditioning may activate potassium ATP channels in humans? *Circulation* **88** (suppl. I), I-569.

Tamura Y, Saito S, Honye J, Hibiya K, Ando T, Moriuchi M, Tsuji M, Mizumura T and Ozawa T. (1990) Cardioprotective effect of nicorandil during percutaneous transluminal coronary angioplasty. *Eur. Heart J.* **11** (suppl.), 52.

Thornton JD, Liu GS, Olsson RA and Downey JM. (1992) Intravenous pretreatment with A_1-selective adenosine analogues protects the heart against infarction. *Circulation* **85,** 659–665.

Tomai F, Crea F, Gaspardone A, Versaci F, Esposito C, Chiariello L and Gioffrè PA. (1993) Mechanisms of cardiac pain during coronary angioplasty. *J. Am. Coll. Cardiol.* **22,** 1892–1896.

Tomai F, Crea F, Gaspardone A, Versaci F, De Paulis R, Penta de Peppo A, Bassano C, Chiariello L and Gioffrè PA. (1994a) Determinants of myocardial ischemia during percutaneous transluminal coronary angioplasty in patients with significant narrowing of a single coronary artery and stable or unstable angina pectoris. *Am. J. Cardiol.* **74,** 1089–1094.

Tomai F, Crea F, Gaspardone A, Versaci F, De Paulis R, Penta de Peppo A, Chiariello L and Gioffrè PA. (1994b) Ischemic preconditioning during coronary angioplasty is prevented by glibenclamide, a selective ATP-sensitive K^+ channel blocker. *Circulation* **90,** 700–705.

Tomai F, Crea F, Gaspardone A, Versaci F, De Paulis R, Castelvecchio S, Gioffrè G, Chiariello L and Gioffrè PA. (1995) Adattamento all'ischemia durante angioplastica coronarica: precondizionamento o reclutamento del circolo collaterale? *Cardiologia* **40** (suppl.), 69.

Tomai F, Crea F, Gaspardone A, Versaci F, De Paulis R, Polisca P, Chiariello L and Gioffrè PA. (1996) Effects of A_1 adenosine receptor blockade by bamiphylline on ischemic preconditioning during coronary angioplasty. *Eur. Heart J.* **17,** 846–853.

University Group Diabetes Program. VIII. (1982) Evaluation of insulin therapy: final report. *Diabetes* **31,** 1–81.

Vaitkus PT, Miller JM, Buxton AE, Josephson ME and Laskey WK. (1994) Ischemia induced changes in human endocardial electrograms during percutaneous transluminal coronary angioplasty. *Am. Heart J.* **127,** 1481-1490.

Van Wylen DGL. (1994) Effect of ischemic preconditioning on interstitial purine metabolite and lactate accumulation during myocardial ischaemia. *Circulation* **89,** 2283–2289.

Walker DM and Yellon DM. (1992) Ischaemic preconditioning: from mechanisms to exploitation. *Cardiovasc. Res.* **26,** 734–739.

Walker DM, Walker JM, Pugsley WB, Pattison CW and Yellon DM. (1995) Preconditioning in isolated superfused human muscle. *J. Mol. Cell. Cardiol.* **27:** 1349–1357.

Wijns W, Serruys PW, Slager CJ, Grimm J, Krayenbuehl HP, Hugenoltz PG and Hess OM. (1986) Effect of coronary occlusion during percutaneous transluminal angioplasty in humans on left ventricular chamber stiffness and regional diastolic pressure–radius relations. *J. Am. Coll. Cardiol.* **7,** 455–463.

Wikstrom BG, Ronquist G and Waldenstrom A. (1995) Dynamics of myocardial metabolism in the preconditioned porcine heart studied using continuous microdialysis. *Eur. Heart J.* **16,** 563–569.

Wohlgelernter D, Cleman M, Highman HA, Fetterman RC, Duncan JS, Zaret BL and Jaffe CC. (1986) Regional myocardial dysfunction during coronary angioplasty: evaluation by two-dimensional echocardiography and 12 lead electrocardiography. *J. Am. Coll. Cardiol.* **7,** 1245–1254.

Yao Z and Gross GJ. (1994) Effects of the K_{ATP} channel opener bimakalin on coronary blood flow, monophasic action potential duration, and infarct size in dogs. *Circulation* **89,** 1769–1775.

9

Preconditioning during cardiac surgery

Louis P. Perrault[1, 2] and Philippe Menasché[1]
[1]Department of Cardiovascular Surgery, Hôpital Lariboisière, 2 rue Ambroise Paré, 75010 Paris, France
[2]Servier Research Institute, 11 rue des Moulineaux, 92150 Suresnes, France

1. Introduction

Endogenous myocardial protection refers to the natural defence mechanisms available to the heart to withstand an ischaemic injury. So far, these mechanisms have been shown to encompass two phenomena most likely interrelated: ischaemic preconditioning (PC) and stress protein synthesis. Ischaemic PC can be defined as the adaptive mechanism induced by a brief period of reversible ischaemia increasing the heart's resistance to a subsequent longer period of ischaemia. Two different time frames are defined for PC, one early (or 'classical' preconditioning) and one late. The therapeutic exploitation of these natural adaptive mechanisms in cardiac surgery is an appealing prospect. PC could be used during aortic cross-clamping to enhance the efficacy of current methods of myocardial protection. However, before PC can be used in patients undergoing heart surgery, at least three major questions have to be answered: Are these endogenous protection mechanisms relevant to surgically induced myocardial ischaemia? How can they be applied during open-heart procedures? How should their efficacy be assessed in cardiac surgical patients?

Ischaemia: preconditioning and adaptation, edited by M.S. Marber and D.M. Yellon

2. Is preconditioning relevant to open-heart surgery?

2.1. Acute protection for global ischaemia

Most of the currently performed surgical procedures involve a transient period of aortic cross-clamping and, consequently, the induction of global ischaemia during which myocardial preservation is ensured by the continuous or intermittent intracoronary delivery of crystalloid or blood-based cardioplegia (CP). A selected review of the experimental and clinical studies that have addressed this situation is presented in *Tables 1–3*. At first glance, the results suggest that the postischaemic recovery of function can be improved by a brief period of global ischaemia preceding the protracted episode of CP arrest. Clinically, a successful use of PC has been reported by Alkhulaifi *et al.* (1994; Yellon *et al.*, 1993) using conservation of adenosine triphosphate (ATP) as the endpoint. Twenty patients were randomized to either PC by two 3-min periods of cross-clamping separated by 2 min of reperfusion prior to an ischaemic insult of 10 min ischaemia and ventricular fibrillation done at a temperature of 37°C. The 10 controls received 10 min cross-clamping with fibrillation without PC. PC caused a significant depletion of ATP content and both PC and control groups showed a significant depletion of ATP at the end of the 10-min ischaemic period. However, PC slowed the rate of ATP depletion so that at the end of the ischaemic insult, the PC group had a significantly higher content than controls (PC 11.5 ± 0.8 vs. control 7.2 ± 0.3 μmol g^{-1} dry weight, $p < 0.005$). In addition to the fact that tissue levels of ATP do not necessarily correlate with an enhanced recovery of function (Cave and Hearse, 1992), the clinical relevance of these findings is hampered by two major caveats.

Reduction of stunning vs. reduction of infarct size. To date, there is convincing evidence that PC improves function by reducing infarct size, not by decreasing stunning of peri-infarct viable myocardium. Unfortunately, not all studies have concomitantly looked at functional indices and markers of myocardial necrosis. The results of those which have adhered to this design show that post-CP function is not improved by PC when there is no evidence of tissue necrosis (Bolling *et al.*, 1994) This may limit the interest of PC in clinical cardiac surgery where postoperative pump failure is usually due to ischaemically induced stunning (Gray *et al.*, 1979; Mazer, 1994).

Table 1. Animal studies

Model	PC	Ischaemia	Reperfusion	Results	Reference
Hypoxic/ischaemic preconditioning in global ischaemia					
Rat	10 min hypoxic	30 min global	30 min	Decreased stunning, reduced cell necrosis	Engelman *et al.*, 1995
Rat	3 x 3 min	30 min global	120–240 min	Less Ca^{2+}-induced Ca^{2+} release, better function	Zucchi *et al.*, 1995
Rat	3 min	21 min global	15 min	Reversible inhibition of mitochondrial ATPase, improved recovery	Vuorinen *et al.*, 1995
Rabbit	10 min hypoxic	30 min global	120 min	Decreased infarction	Walsh *et al.*, 1995
Rabbit	5 min	35 min	10 min	No improvement in infarct or function	Quantz *et al.*, 1994
Pharmacological preconditioning in global ischaemia					
Guinea pig	90 sec–5 min PC aprikalim	30 min	60 min	Synergistic benefit	Hecquet *et al.*, 1994
Mouse	5 min PC adenosine pinacidil	60 min	60 min	Less leukocyte adhesion and migration	Akimitsu *et al.*, 1994
Rat	5 min PC	20 min adenosine-free	35 min	No PC with adenosine	Cave *et al.*, 1991
Rat	4 x 5 min vs. $MgCl_2$ arrest	30 min global	30 min	Improved function, less intracellular Na^+, Ca^{2+}, H^+. No difference with $MgCl_2$	Steenbergen *et al.*, 1993
Rat	RP52891 (PCO)	25 min	30 min	Improved function	Grover *et al.*, 1990

(*continued*)

(**Table 1** *continued*)

Model	PC	Ischaemia	Reperfusion	Results	Reference
Rat	Glibenclamide 5 min PC	25 min	30 min	Improved function, inhibited by glibenclamide	McCullough *et al.*, 1991
Rabbit	5 min PC, BK + glyburide	30 min	10 min	BK_1 receptor stimulation via K_{atp} channels	Wall *et al.*, 1994
Rabbit	5 min PC vs. BK	30 min	120 min	Reduced infarct size	Wall *et al.*, 1994
Rabbit	Aprikalim	20 min	120 min	Improved recovery	Cohen *et al.*, 1993
Rabbit	2 min PC	30 min	5 min reperf. + 72 h reperf. dypiridamole	Inhibition of nucleotide potentiates PC reduced infarct size	Itoya *et al.*, 1992
Rabbit	2 min PC, acadesine	30 min	180 min	PC and acadesine improved function	Tsuchida *et al.*, 1992
Rabbit	5 min PC, CCPA infusion infusion up 72 h	30 min regional	180 min	Desensitization to CCPA, PC not protective when desensitized, reduced infarct size	Tsuchida *et al.*, 1994
Rabbit	Glucose-free 15 min	30 min regional	120–180 min glucose free	Reduced infarct size, blocked by PKC inhibitors	Goto *et al.*, 1995
Dog	Aprikalim	15 min reg	3 h	Less infarct, blocked by glibenclamide	Auchampach *et al.*, 1992
Swine	Nicorandil, ISDN	60 min regional	180 min	Improved function, antineutro-phil activity	Galiè *et al.*, 1995
Swine	10 x 2 min SPT±CCPA	Repetitive occlusions	2 min	No protection from stunning by adenosine	Tang *et al.*, 1994

BK, bradykinin; BK_1, bradykinin 1 receptor; CCPA, 2-chloro-*N*-cyclopentyladenosine; ISDN, isosorbide dinitrate; PCO, potassium channel opener; SPT, 8-*p*-sulphophenyl-theophylline.

Table 2. Human studies

Model	PC	Ischaemia	Reperfusion	Results	Reference
Ischaemic preconditioning					
CABG	3 min PC, 2 min reperf. repeated x 1	10 min (37°C)	—	Enhanced conservation of ATP	Yellon *et al.*, 1993 Alkhulaifi *et al.*, 1994
CABG	3 min PC, 2 min reperf. repeated x 1	10 min 32°C vs. 37°C	—	No ATP differences at hypothermia	Di Salvo *et al.*, 1995
Thrombolysis	Prodromal angina	—	—	Smaller infarct size	Ottani *et al.*, 1995
Pharmacological preconditioning					
CABG	Adenosine PC	Not stated	Not stated	Improved haemodynamics	Lee *et al.*, 1994
PTCA	90 sec ischaemia	—	180 sec aminophylline	Aminophylline abolishes PC	Claeys *et al.*, 1994
PTCA	Glibenclamide	2 x 2 min balloon inflation	—	Lower ST segment shift	Tomai *et al.*, 1994
LV muscle	A1+P1 receptor stimulation	30 min hypoxia	60 min reox.	Enhanced function	Wollmering *et al.*, 1994
Myocyte	90 min	30 min	—	Reduced cell injury	Ikonomidis *et al.*, 1994
Atrial trabeculae	3 min pacing	90 min hypoxia	120 min	Improved recovery	Walker *et al.*, 1995
Atrial trabeculae	3 min PC	90 min	120 min	Improved recovery, PKC activation	Speechly-Dick *et al.*, 1995

CABG, coronary artery bypass grafting; LV, left ventricular; PTCA, percutaneous transluminal coronary angioplasty.

Table 3. Preconditioning and cardioplegia

Model	PC	Ischaemia	Results	Reference
Guinea pig	Nicorandil	120 min STH cardioplegic arrest (37°C)	Improved force–frequency relationship, improved function	Sugimoto *et al.*, 1995a Sugimoto *et al.*, 1995b Sugimoto *et al.*, 1994
Rat	Cromakalim BMS 180048 (PCO)	30 min at 37°C 150 min at 0°C	Improved function	Grover and Sleph, 1995
Rat	10 min hypoxic	STH 4–6 h (4°C)	Improved function	Engelman *et al.*, 1994
Rat	5 min PC 5 min PC	10–25 min (37°C) 45–160 min (20°C)	Improved function HEP unchanged	Cave and Hearse, 1992b
Rat	5 min PC	STH cardioplegia, 120–200 min arrest (20°C)	PC benefit additive to cardioplegia and hypothermia, improved function	Cave and Hearse, 1992a
Rat	3 min PC	STH cardioplegia, 35 min ischaemia blood perfused (37°C)	More ischaemic contracture with PC, less acidosis, no difference in recovery	Kolocassides *et al.*, 1995
Rat	5 min PC	STH cardioplegia 30 min (37°C)	PC improves function if cardioplegia delivery impaired	Galinanes *et al.*, 1995
Rat	3 min PC	35 min global, STH (37°C)	No additive protection from PC + cardioplegia	Kolocassides *et al.*, 1994
Rabbit	1–5 min ischaemia	2.5 h cardioplegia (15°C)	Improved function	Illes *et al.*, 1994
Rabbit	5–45 min	120 min cardioplegia STH (34°C)	No improved recovery if pre-arrest stunning	Bolling *et al.*, 1994
Dog	Glucose insulin K^+	60 min cardioplegia on bypass (37°C)	Improved function	Nishida *et al.*, 1993
Dog	42°C x 15 min on bypass	15 min regional 3 h reperf. (37°C)	Improved function	Robinson *et al.*, 1995
Swine	15 min heat shock 42°C	2 h cardioplegia (4–6°C), 1 h reperf. (37°C)	Improved function increased SOD	Liu *et al.*, 1992
Swine	60 min LAD occlusion	60 min cardioplegic arrest on bypass, 1 h reperf. (37°C)	Improved function	Maulik *et al.*, 1995

PCO, potassium channel opener; St Thomas' Hospital solution; SOD, superoxide dismutase.

Redundancy between CP and PC. Regardless of the precise mechanism by which PC exerts its cardioprotective effects, the final consequence of the intracellular signalling pathway seems to involve a slowing of the decay of high energy phosphates (HEP) during the prolonged period of ischaemia at least when there is some residual collateral blood flow (Kolocassides, 1995). It is therefore conceivable that PC may enhance the effects of a technique like normothermic ventricular fibrillation – the one used in Alkhulaifi's study – which is expected to cause a drastic decline in ATP levels (Alkhulaifi *et al.*, 1994). Likewise, the favourable outcomes of patients undergoing coronary artery bypass grafting (CABG) under intermittent aortic cross-clamping during which the heart is usually fibrillating might be explained by the PC effect of the initial episode of aortic occlusion. Conversely, the ATP-sparing effect of PC may become redundant to that of hypothermia and CP, which remain the most commonly used methods of myocardial preservation. Support for this hypothesis comes from the observation that the benefits of PC seen with normothermic ventricular fibrillation are lost when the temperature is lowered to 32°C, most likely because this mild degree of hypothermia allows a substantial preservation of HEPs. Experimentally, PC and CP have been found to be equally protective up to 35 min of global ischaemia at 37°C but without additive effects. Interestingly, PC improved protection above that afforded by CP only when the ischaemia period was prolonged for more than 120 min or when distribution of CP was impaired by proximal blockade of coronary arteries. We suspect that in these two situations, the benefits of PC are explained by a reduction in the amount of necrosis that may have been caused by suboptimal CP protection. Although the influence of the temperature of the CP solution on the beneficial effects of PC has not been studied as such, Cave *et al.* showed improved recovery after a 5 min ischaemic PC period up to 20 min of global ischaemia at 37°C and up to 160 min at 20°C with no changes in HEP (Cave and Hearse, 1991, 1992a) but interestingly found no consistent correlation with tissue content of HEP and enhanced ischaemic recovery (Cave, 1992b).

Thus, in the setting of standard CP protection where attempts are made at both minimizing the length of the ischaemic insult and optimizing the delivery of CP, the added benefits of a prearrest PC stimulus remain elusive. The question is to know whether this pre-CP ischaemic episode may even be harmful. Although this issue is as yet unsettled, the results of a recent study conducted in our institution lead us to be cautious (see below).

Acute protection in regional ischaemia. Most of the experimental studies on PC have used models of regional ischaemia and infarct size as the major endpoint. Consequently, the technique might be better suited for operations occasionally performed in the beating heart, with or without cardiopulmonary bypass (CPB), and during which the completion of the distal anastomosis requires elective isolation of the coronary vessel targeted for grafting. It remains, however, to be determined whether ischaemic tolerance to this coronary clamping could be improved by a prior brief episode of vessel occlusion. The soundness of this hypothesis is supported by the finding that patients undergoing percutaneous transluminal coronary angioplasty (PTCA) experience fewer signs of ischaemia during the second balloon inflation compared to the first (Deutsch *et al.*, 1990).

3. How could PC be implemented in clinical practice?

For years, surgeons have achieved excellent results using brief periods (< 10 min and < 30 min cumulative) of coronary occlusion with intermittent reperfusion during sequential construction of proximal anastomoses for coronary artery bypass surgery (Bonchek *et al.*, 1992; Flameng *et al.*, 1984). These outcomes may be related to the PC effect of the initial period of occlusion and to the absence of cumulative stunning that may occur after multiple repetitive coronary occlusion as shown recently by Wechsler's group (Abd-Elfattah *et al.*, 1995). Also, the recent interest in minimally invasive coronary bypass surgery with or without circulatory support (Subramanian, 1995) and revascularization without aortic cross-clamping (Pirk and Kellovsky, 1995) may result in situations where ischaemic PC could enhance myocardial preservation. The regional ischaemia produced by occlusion of the target coronary artery for construction of the anastomosis may be used as a PC stimulus.

The preceding caveats do not intend to deter from the use of PC in clinical cardiac surgery. They rather emphasize that, at present, extrapolation of standard ischaemic PC regimens to the surgical setting may be hazardous. The application of these regimens would be further complicated by the uncertainty regarding the optimal duration of the PC stimulus and of the intervening reperfusion. Nevertheless, the exploitation of the heart's endogenous defence mechanisms appears a logical prospect. It is likely, however, that a more effective approach would be the identification of the pharmacological mediators of PC in order to use them therapeutically. Based on the current knowledge of the mechanisms of PC, three types of targets can be considered.

3.1. Triggers

Adenosine. Amongst the prime candidates for PC agent is adenosine. Strategies designed to stimulate adenosine receptors, to provide exogenous adenosine or to enhance endogenous adenosine production have been used to precondition. In addition to the numerous experimental studies that have demonstrated the role of A1 receptor stimulation in the PC effect, clinical trials using CP plus acadesine, an adenosine regulating agent, have shown benefits in terms of clinical outcome (Menasché *et al.*, 1995).

Some studies have found little role for adenosine in PC which may be species-related (Lasley and Mentzer, 1993) and no protection afforded by adenosine against stunning (Tang *et al.*, 1994). Conversely, Lee found that infusion of adenosine prior to bypass improved postoperative ventricular function through what they believed was a PC effect (Lee *et al.*, 1994). *In vitro* studies using human left ventricular muscle confirmed a role for A1 receptor stimulation in enhancing post-ischaemic function (Wollmering *et al.*, 1994), which is also consistent with the finding that aminophylline abolishes the PC induced during percutaneous transluminal coronary angioplasty (PTCA) (Claeys *et al.*, 1994).

Alpha-adrenergic receptor agonists. Mitchell *et al.* have identified a role for alpha 1-adrenergic receptor stimulation in reproducing a PC effect by using a 2 min perfusion of phenylephrine (Banerjee *et al.*, 1993; Mitchell *et al.*, 1993, 1995; Winter *et al.*, 1992). The alpha 1-mediated PC causes protein kinase C (PKC) translocation (Mitchell *et al.*, 1995). In the study by Nattel and Hu in isolated perfused rat hearts, alpha 1b-receptor blockade prevented PC while there was no effect from alpha 1a blockade (Hu and Nattel, 1995). The clinical applicability of this approach remains to be determined. Interestingly, there is recent evidence that CPB alone can elicit PC (Burns *et al.*, 1995), possibly through the activation of both adenosine and alpha-adrenergic receptors as the PC effect was abolished by blockade of these receptors.

Beta-adrenergic, muscarinic and bradykinin receptor agonists. Beta-adrenergic muscarinic and bradykinin (Brew *et al.*, 1995; Wall *et al.*, 1994) receptor activation is also thought to trigger the intracellular signalling pathway which underlies the cardioprotective effects of PC and represents an alternative approach worthy of investigation.

3.2. Mediators

PKC activators. Much experimental evidence points to PKC as a key mediator of the signalling pathway activated by PC. A role for PKC activation in mediation of the PC effect has been proven by use of both antagonists such as polymixin and staurosporin and stimulators such as phorbol ester (Mitchell *et al.*, 1995; Ytrehus *et al.*, 1994). Ikonimidis has confirmed the role of PKC activation in both ischaemic and adenosine-induced PC in human cardiomyocytes (Ikonomidis *et al.*, 1995). However, currently available activators of PKC have tumour-promoting effects that make them useful as tools rather than as drugs for potential human use.

3.3. Effectors

ATP-sensitive potassium channel openers. The most studied effector of PC has been the ATP-sensitive potassium (K_{ATP}) channel identified in dog (Auchampach *et al.*, 1992; Gross and Auchampach, 1992; Grover *et al.*, 1992; Yao and Gross, 1993, 1994), rat (Grover *et al.*, 1990; Liu and Downey, 1992; McCullough, 1991), rabbit (Cohen *et al.*, 1993) and man (Tomai *et al.*, 1994). Potassium channel opening occurs at the time of ischaemia and causes a shortening of the action potential duration in the myocardial cell which may prevent the development of stunning (Shigematsu *et al.*, 1995). The specific agents studied include nicorandil, aprikalim, pinacidil and RP 52891. Potassium channel openers may have an effect through inhibition of leukocyte adhesion and migration (Akimitsu *et al.*, 1994; Galiè *et al.*, 1995). Some vasodilatory effect or beneficial event at the level of the microcirculation may also occur, leading to better substrate delivery during reperfusion .

The possibility of duplicating the cardioprotective effects of ischaemic PC by potassium channels openers has been demonstrated by Sugimoto in isolated guinea pig papillary muscle subjected to CP arrest (Sugimoto *et al.*, 1994, 1995a,b). Likewise, we have demonstrated that ischaemic and nicorandil PC equally improved recovery following normothermic (Menasché, 1996) or mildly hypothermic potassium-based CP arrest (Grover and Sleph, 1995). The relevance of this mechanism to the human species is further supported by the finding that glibenclamide prevents patients undergoing PTCA acquiring tolerance to sequential balloon-induced ischaemia (Tomai *et al.*, 1994). Additional evidence for the involvement of K_{ATP} channels as end-effector of the PKC mediated signalling pathway elicited by PC in human comes from the recent study from Yellon's group using right atrial trabeculae (Speechly-Dick *et al.*, 1995).

Na^+/H^+ exchangers inhibitors. Na^+/H^+ exchange is an important mechanism for cell death in myocardial ischaemia and reperfusion (Klein *et al.*, 1995). PC may lead to stimulation of the Na-K-2Cl cotransporter, and inhibition of the Na^+/H^+ exchanger. These effects may prevent an excessive rise in intracellular sodium content (Ramasamy *et al.*, 1995) contributing to the prevention of neutrophil-mediated reperfusion injury (Faes *et al.*, 1995), and to the development of intracellular acidosis and calcium overload (Kida *et al.*, 1991; Steenbergen *et al.*, 1993). Inhibition of this exchanger might duplicate the cardioprotective effects of ischaemic PC and be useful clinically, although this remains to be determined.

3.4. Delayed protection

An alternative means of exploiting the heart's own defence mechanisms could be to stimulate the synthesis of stress proteins, which have been shown to have cardioprotective effects (Marber *et al.*, 1993) as evidenced by a decrease in infarct size and an attenuation of postischaemic stunning (Sun *et al.*, 1995). This approach is supported by the report that hypoxic PC or hyperthermic bypass (Liu *et al.*, 1992) can induce the expression of stress proteins within a relatively short time frame consistent with that of a surgical procedure. These results, however, should be interpreted in light of two caveats: except for one study, they demonstrate an increased expression of mRNAs, not of the relevant proteins, and the time course required for these proteins to be operative may exceed that of the early postbypass period when optimal protection is needed the most. In addition, these results have been obtained in normal hearts and may not be readily extrapolated to diseased hearts, where patterns of stress protein synthesis may differ. This hypothesis is supported by the finding that the myocardial content of hsp72 in patients undergoing various open-heart procedures remains unaltered thoughout the period of CPB (McGrath *et al.*, 1995) when assayed 15 min after bypass weaning. A reperfusion component of HSP expression (Heads *et al.*, 1995) may be important for stimulation of *HSP70* mRNA expression whereas *HSP60* mRNA is expressed readily. mRNAs for *HSP70* increased during the experiment but there was no induction of hsp70i protein expression over the entire duration of the protocol used (PC followed by 30 min of ischaemia and 2 h of reperfusion). The hsp70i protein levels were significantly less after 2.5 h of experiment in the PC protocol than levels measured in heat stressed ventricles and the ones measured 24 h following PC.

3.5 A word of caution on the application of PC to clinical bypass surgery

Increased ischaemia. In the same way that the metabolic benefit of PC (higher content of high-energy phosphates) may be lost when lowering bypass temperature from 37°C to 32°C, there may be circumstances under which ischaemic PC could be detrimental. In a pilot study performed in our centre, 20 case-matched patients undergoing CABG were randomized to either normothermic continuous retrograde CP or ischaemic PC with 3 min of aortic cross-clamping with 2 min of reperfusion before the onset of CP arrest. There were no differences in pre- or intraoperative variables between the two groups. At the end of arrest (72 ± 6 and 64 ± 5 min in the control and PC groups respectively [mean ± SEM], the release of CK-MB from the myocardium (calculated as the difference between coronary sinus and radial artery values) was markedly greater than in controls (5.7 ± 1.7 ng ml^{-1} vs. 1.9 ± 1.1 ng ml^{-1}, p=0.05). The transmyocardial lactate gradient was shifted towards production in the PC group (+0.22 ± 0.13 mmol l^{-1}) and towards extraction in the control group (–0.06 ± 0.21 mmol l^{-1}). These results were not confounded by the use of aprotinin, which has recently been reported to negate the protective effects of PC, since the administration of this drug was evenly distributed among the two groups. Although there were no PC-related adverse clinical events, this trend towards an increased initial ischaemic insult should not be dismissed.

Use of aprotinin. The serine protease inhibitor aprotinin, used extensively in clinical practice to reduce postoperative bleeding, has been reported to cause increased myocardial infarction in certain series (Cosgrove *et al.*, 1992). In a recent experimental study performed in fully heparinized sheep, aprotinin increased infarct size after a 60 min regional occlusion compared to control (Bukhari *et al.*, 1995). It also attenuated the protective effect of PC on infarct size. The attenuation of PC benefit in this setting emphasizes the need to precisely establish the safety and indications of ischaemic PC in clinical practice and to control factors such as premedication (allopurinol, β-adrenergic blockade), use of β-adrenergic inotropes, phosphodiesterase inhibitors (aminophylline) and concurrent medication (aprotinin).

4. How to assess the protective effects of ischaemic PC and stress protein synthesis in clinical practice?

4.1. Biochemical and enzymatic measurements

Given the excellent results achieved with current CP and non-CP techniques, it is unlikely that standard biochemical or enzymatic assays such as measurements of lactate release, of the myocardial isoform of creatinine phosphokinase or troponin release measurements, will be sensitive enough to detect improvements resulting from PC regimens. However, some information may be drawn from these simple and readily available tests, that may detect, as was the case in our pilot series, a trend toward an increased ischaemic insult in the PC group.

4.2. ATP measurements

Measurement of myocardial ATP content has been useful in some series to express the metabolic endpoints of techniques, such as PC, designed to improve protection. The significance of preserved ATP stores is, as yet, not fully clarified as improvement in recovery of function may occur independently of ATP content. Lastly, the results from this endpoint should be interpreted in the light of other confounding factors such as temperature. The superior ATP conservation in the PC group in the initial Yellon series did not lead to any clinical benefit (Alkhulaifi *et al.*, 1994). Would one modify the protection technique on the basis of such meager evidence? It is unlikely.

Proto-oncogene and cardiac gene expression studies may be useful for uncovering the precise mechanisms through which PC protects the heart, and remain tools for research. Studies must first show direct evidence that changes in genetic make-up are associated with protection and that inhibition of these changes prevents the PC benefit. In the case of mRNA measurements, the information obtained should ideally be corroborated with the expression of the proteins (as in the case of stress protein) to link the change in genetic message to effective protein synthesis. Some experimental studies have addressed these issues (Heads *et al.*, 1995; Liu *et al.*, 1992). Likewise, in a model of hypoxic PC, improvement of function has been shown to be associated with an increased expression of hsp70 mRNAs (Liu *et al.*, 1992). Our preliminary experience with ischaemic PC in the clinical setting has not shown any difference in atrial c-*fos* or hsp70 mRNA expression between PC (2 min global ischaemia and 3 min reperfusion before continuous retrograde normothermic CP) and control (CP only). This may be related to high initial levels of HSP in patients undergoing cardiac surgery (McGrath *et al.*, 1995) or to the effect of CP as demonstrated by Aebert *et al.* (1995).

Molecular biology techniques may be more useful to uncover the mechanisms involved in PC than to evaluate its efficacy. Numerous factors interplay in patients undergoing cardiac surgery such as the underlying disease, the concurrent medication and the use of bypass. These factors may render proper interpretation of data difficult.

Ultimately, clinically useful criteria will have to be studied to determine the efficacy of PC regimens, which patients will benefit most and which myocardial protection strategies are most appropriate. In the end, no matter what the sensitive indices of cellular damage tell us, benefit should be measured by clinical endpoints such as improved immediate or late myocardial function or perhaps even improved survival. Any benefit is likely to be greatest in certain subgroups of patients such as those with problematic CP delivery or poor ventricular function. These considerations must be addressed before recommending the routine application of techniques or pharmacological interventions that bear potential side-effects, and may even be ineffective. Hence the need for caution in using PC in cardiac surgery at this point in time. Until then, continued efforts must be made to try to understand and harness, for the benefit of our patients, the immense endogenous protective mechanisms within the heart.

Acknowledgements

Dr Perrault is supported by the Clinician-Scientist program of the Medical Research Council of Canada.

References

Abd-Elfattah AS, Ding M and Wechsler AS. (1995) Intermittent aortic cross-clamping prevents cumulative adenosine triphosphate depletion, ventricular fibrillation, and dysfunction (stunning): is it preconditioning? *J. Thorac. Cardiovasc. Surg.* **110,** 328–329.

Aebert H, Schunkert H, Ehr T, Cornelius T, Preuner J, Riegger GAJ and Birnbaum DE. (1995) Induction of heat shock protein and protooncogenes following cardioplegia in murine and human hearts. *Thorac. Cardiovasc. Surgeon* **43** (suppl. I), 148.

Akimitsu T, Gute DC and Korthuis RJ. (1994) Ischemic preconditioning attenuates postischemic leukocyte adhesion and emigration: role of adenosine and ATP-sensitive potassium channels. *Circulation* **90** (suppl. I), A2561.

Alkhulaifi AM, Yellon DM and Pugsley WB. (1994) Preconditioning the human heart during aorto-coronary bypass surgery. *Eur. J. Cardiothorac. Surg.* **8,** 270–276.

Auchampach JA, Grover GJ and Gross GJ. (1992) Blockade of ischaemic preconditioning in dogs by the novel ATP-dependent potassium channel antagonist sodium 5-hydroxydecanoate. *Cardiovasc. Res.* **26,** 1054–1062.

Auchampach JA, Maruyama M, Cavero I and Gross GJ. (1992) Pharmacological evidence for a role of ATP-dependent potassium channels in myocardial stunning. *Circulation* **86,** 311–319.

Banerjee A, Winter-Locke C, Rogers KB, Mitchell MB, Brew EC, Cairns CB, Bensard DD and Harken AH. (1993) Preconditioning against myocardial dysfunction after ischemia and reperfusion by an α1-adrenergic mechanism. *Circ. Res.* **73,** 656–670.

Bolling SF, Olszanski DA, Childs KF, Gallagher KP and Ning XH. (1994) Stunning, preconditioning, and functional recovery after global myocardial ischemia. *Ann. Thorac. Surg.* **58,** 822–827.

Bonchek LI, Burligame MW, Vazales BE, Lundy EF and Gassman CJ. (1992) Applicability of noncardioplegic coronary bypass to high-risk patients. *J. Thorac. Cardiovasc. Surg.* **103,** 230–237.

Brew EC, Mitchell MB, Rehring TF, Gamboni-Robertson F, McIntyre RC, Harken AH and Banerjee A. (1995) Role of bradykinin in cardiac functional protection after global ischemia–reperfusion in rat heart. *Am. J. Physiol.* **269,** H1370–H1378.

Bukhari EA, Krukenkamp IB, Burns PG, Gaudette GR, Schulman J, Al-Fagih MR and Levitsky S. (1995) Does aprotinin increase the myocardial damage in the setting of ischemia and preconditioning? *Ann. Thorac. Surg.* **60,** 307–310.

Burns PG, Krukenkamp IB, Caldarone CA, Gaudette GR, Bukhari EA and Levitsky S. (1995) Does cardiopulmonary bypass alone elicit myoprotective preconditioning? *Circulation* **92** (suppl. II), II447–II451.

Cave AC and Hearse DJ. (1991) Preconditioning in the globally ischaemic rat heart: influence of duration and temperature of ischemia. *J. Mol. Cell. Cardiol.* **23** (suppl. V), S73.

Cave AC and Hearse DJ. (1992b) Ischaemic preconditioning and contractile function: studies with normothermic and hypothermic global ischaemia. *J. Mol. Cell. Cardiol.* **24,** 1113–1123.

Cave AC and Hearse DJ. (1992a) Ischemic preconditioning enhances post-ischemic function and reduces creatine kinase leakage in the rat heart even when used in conjunction with hypothermic cardioplegia. *Circulation* **865** (suppl, I), I81.

Cave AC, Downey JM and Hearse DJ. (1991) Adenosine fails to substitute for preconditioning in the globally ischemic isolated heart. *J. Mol. Cell. Cardiol.* **23** (suppl. III), S76 (abstract).

Claeys MJ, Vrints CJ, Bosmans JM, Conraads VM and Snoeck JP. (1994) Aminophylline abolishes ischemic preconditioning during angioplasty. *Circulation* **90** (suppl. I), I-477.

Cohen NM, Wise RM, Wechsler AS and Damiano RJ. (1993) Elective cardiac arrest with a hyperpolarizing adenosine triphosphate-sensitive potassium channel opener; a novel form of myocardial protection. *J. Thorac. Cardiovasc. Surg.* **106,** 317–328.

Cosgrove DM III, Heric B and Lyttle B, *et al.* (1992) Aprotinin therapy for reoperative myocardial revascularization: a placebo-controlled study. *Ann. Thorac. Surg.* **54,** 1031–1038.

Deutsch E, Berger M, Kussmaul WG, Hirshfield JW, Herrman HC and Laskey WL. (1990) Adaptation to ischemia during percutaneous transluminal coronary angioplasty. Clinical, hemodynamic, and metabolic features. *Circulation* **82,** 2044–2051.

Di Salvo C, Hemming A, Jenkins D, Yellon D, Oakley RE, Wright J, Walesby R and Feneck R. (1995) Can the human myocardium be preconditioned with ischaemia under hypothermic conditions? *Proceedings of the 9th Annual Meeting of the European Association for Cardiothoracic Surgery,* Paris, 25–27 September 1995, p. 324.

Engelman DT, Watanabe M, Engelman R, Rousou JA, Kisin E, Kagan VE, Maulik N and Das DK. (1995) Hypoxic preconditioning preserves antioxidant reserve in the working rat heart. *Cardiovasc. Res.* **29,** 133–140.

Engelman DT, Chen CZ, Watanabe M, Engelman R, Rousou JA, Flack JE, Deaton DW, Cordis GA and Das DK. (1994) Improved four and six hour myocardial preservation by hypoxic preconditioning. *Circulation* **90** (suppl. I), I-46.

Faes FC, Sawa Y, Ichikawa H, Shimazaki Y, Ohashi T, Fukuda H, Shirakura R and Matsuda H. (1995) Inhibition of Na^+/H^+ exchanger attenuates neutrophil-mediated reperfusion injury. *Ann. Thorac. Surg.* **60,** 377–381.

Flameng W, Van der Vusse GJ, De Meyere R, Borgers M, Sergeant P, Vander Meersch E, Geboers and Suy R. (1984) Intermittent aortic cross-clamping versus St. Thomas' Hospital cardioplegia in extensive aorta-coronary bypass grafting. *J. Thorac. Cardiovasc. Surg.* **88,** 164–173.

Galiè N, Guarnieri C, Ussia GP, Zinmaro M, Traini AM, Parlangeli R, Vaona I, Branzi A and Magnani B. (1995) Limitation of myocardial infarct size by nicorandil after sustained ischemia in pigs. *J. Cardiovasc. Pharmacol.* **26,** 477–484..

Galinanes M, Argano V and Hearse DJ. (1995) Can ischemic preconditioning ensure optimal myocardial protection when delivery of cardioplegia is impaired? *Circulation* **92** (suppl. I), II-389–II394.

Goto M, Tsuchida A, Liu Y, Cohen MV and Downey JM. (1995) Transient inhibition of glucose uptake mimics ischemic preconditioning by salvaging ischemic myocardium in the rabbit heart. *J. Mol. Cell. Cardiol.* **27,** 1883–1894.

Gray R, Maddhai J, Berman D, *et al.* (1979) Scintigraphic and hemodynamic demonstration of transient left ventricular dysfunction immediately after uncomplicated coronary artery bypass grafting. *J. Thorac. Cardiovasc. Surg.* **77,** 504–510.

Gross GJ and Auchampach JA. (1992) Blockade of ATP-sensitive potassium channels prevents myocardial preconditioning in dogs. *Circ. Res.* **70,** 223–233.

Grover GJ, Dzwonczyk S and Sleph PG. (1990) Reduction of ischemic damage in isolated rat hearts by the potassium channel opener RP 52891. *Eur. J. Pharmacol.* **191,** 11–18.

Grover GJ, Sleph PG and Dzwonczyk S. (1992) Role of myocardial ATP-sensitive potassium channels in mediating preconditioning in the dog heart and their possible interaction with adenosine A1-receptors. *Circulation* **86,** 1310–1316.

Grover GJ and Sleph PG. (1995) Protective effect of K_{atp} openers in ischemic rat heart treated with a potassium cardioplegic solution. *J. Cardiovasc. Pharmacol.* **26,** 698–706.

Heads RJ, Latchman DS and Yellon DM. (1995) Differential stress protein mRNA expression during early ischaemic preconditioning in the rabbit heart and its relationship to adenosine receptor function. *J. Mol. Cell. Cardiol.* **27,** 2133–2148.

Hecquet C, Mestre M and Cavero I. (1994) Synergistic cardioprotective effect of a subtreshold concentration of aprikalim and a subthreshold ischemic preconditioning stress. *Circulation* **90** (suppl. I), I-48.

Hu K and Nattel S. (1995) Mechanisms of ischemic preconditioning in rat hearts. Involvement of α1B-adrenoreceptors, pertussis toxin-sensitive G-proteins, and protein kinase C. *Circulation* **92,** 2259–2265.

Ikonomidis JS, Shirai T, Weisel RD, Derylo B, Rao V, Whiteside CI, Mickle DAG and Li RK. (1995) "Ischemic" or adenosine preconditioning of human ventricular cardiomyocyte is protein C dependent. *Circulation* **92** (suppl. I), I-12.

Ikonomidis JS, Shirski T, Weisel RD and Mickle DAG. (1994) Human cardiomyocyte preconditioning is adenosine A1-receptor dependent. *Circulation* **90** (suppl. I), I-477.

Illes RW, Wright JK, Inners-McBride K, Yang CJ and Tristan A. (1994) Ischemic preconditioning improves preservation with crystalloid cardioplegia. *Ann. Thorac. Surg.* **58,** 1481–1485.

Itoya M, Kazuyuki T, Ishimoto R and Iimura O. (1992) Nucleotide transport inhibitors enhanced the infarct size-limiting effect of preconditioning. *Circulation* **86** (suppl. I), I25.

Kida M, Fujiwara H, Ishida M, Kawai C, Ohura M, Miura I and Yabuuchi. (1991) Ischemic preconditioning preserves creatine phosphate and intracellular pH. *Circulation* **84,** 2495–2503.

Klein HH, Pich S, Bohle RM, Wollenweber J and Nebendahl K. (1995) Myocardial protection by Na^+–H^+ exchange inhibition in ischemic reperfused porcine hearts. *Circulation* **92,** 912–917.

Kolocassides KG, Galinanes M and Hearse DJ. (1994) Ischemic preconditioning, cardioplegia or both? *J. Mol. Cell. Cardiol.* **26,** 1411–1414.

Kolocassides KG, Galinanes M and Hearse DJ. (1995) Preconditioning accelerates contracture and ATP depletion in blood-perfused rat hearts. *Am. J. Physiol.* **269,** H1415–H1420.

Lasley RD and Mentzer RM. (1993) Adenosine increases lactate release and delays onset of contracture during global low flow ischemia. *Cardiovasc. Res.* **27,** 96–101.

Lee HT, Sanchala VT, Reed GE, Belkin RN and LaFaro RJ. (1994) Preconditioning the human myocardium with adenosine during cardiac surgery. *Circulation* **90** (suppl. I), I-48.

Liu X, Engelman RM, Moraru II, Rousou JA , Flack JE III, Deaton DW, Maulik N and Das DK. (1992) Heat shock. A new approach for myocardial preservation in cardiac surgery. *Circulation* **86** (suppl. II), II-358–II-363.

Liu Y and Downey JM. (1992) Ischemic preconditioning protects against infarction in rat heart. *Am. J. Physiol.* **263,** H1107–H1112.

Marber MS, Latchman DS, Walker JM and Yellon DM. (1993) Cardiac stress protein elevation 24 hours after brief ischemia or heat stress is associated with resistance to myocardial infarction. *Circulation* **88,** 1264–1272.

Maulik N, Engelman RM, Wei Liu X, Rousou JA, Flack JE, Deaton DW and Das DK. (1995) Drug-induced heat-shock preconditioning improves postischemic ventricular recovery after cardiopulmonary bypass. *Circulation* **92** (suppl.II), II381–II388.

Mazer CD. (1994) Stunned myocardium in warm and cold surgery. *J. Card. Surg.* **9** (suppl. 3), 503–505.

McGrath LB, Locke M, Cane M, Chen C and Ianuzzo CD. (1995) Heat shock protein (HSP 72) expression in patients undergoing cardiac operations. *J. Thorac. Cardiovasc. Surg.* **109,** 370–376.

Menasché P, Jamieson WRE, Flameng W and Davis MK. (1995) Acadesine: a new drug that may improve myocardial protection in coronary artery bypass grafting (CABG). *J. Thorac. Cardiovasc. Surg.* **110,** 1096–1106.

Menasché P, Kevelaïtis E, Mouas C, Grousset C, Piwnica A and Bloch G. (1995) Preconditioning with potassium channel openers: a new concept for enhancing cardioplegic protection? *J. Thorac. Cardiovasc. Surg.* **110,** 1606–1614.

Mitchell MB, Winter CB, Locke-Winter CR, Banerjee A and Harken AH. (1993) Cardiac preconditioning does not require myocardial stunning. *Ann. Thorac. Surg.* **55,** 395–400.

Mitchell MB, Meng X, Ao L, Brown JM, Harken AH and Banerjee A. (1995) Preconditioning of isolated rat heart is mediated by protein kinase C. *Circ. Res.* **76,** 73–81.

McCullough JR, Normandin DE, Conder ML, Sleph PG, Dzwonczyk S and Grover GJ. (1991) Specific block of the anti-ischemic actions of cromakalim by sodium 5-hydroxydecanoate. *Circ. Res.* **69,** 949–958.

Nishida H, Grooters RK, Thieman KC, Soltanzadeh H, Schneider RF and Merkley DF. (1993) Enhanced ventricular recovery from high dose glucose, insulin and potassium with cardiopulmonary bypass support prior to cardioplegic arrest. *Eur. J. Cardiothorac. Surg.* **7,** 409–413.

Ottani F, Galvani M, Ferrini D, Sorbello F, Limonetti P, Pantioli D and Rusticali F. (1995) Prodromal angina limits infarct size: a role for ischemic preconditioning. *Circulation* **91,** 291–297.

Pirk J and Kellovsky P. (1995) An alternative to cardioplegia. *Ann. Thorac. Surg.* **60,** 464–465.

Quantz M, Carsley L, Shum-Tim D, Tchervenkov C and Chiu RCJ. (1994) The enigma of myocardial preconditioning models. *J. Card. Surg.* **9** (suppl. 9), 532–536.

Ramasamy R, Liu H, Lundmark J and Schaefer S. (1995) Ischemic preconditioning stimulates sodium and proton transport in isolated rat hearts. *J. Clin. Invest.* **96,** 1464–1472.

Robinson BL, Morita T, Toft DO and Morris JJ. (1995) Accelerated recovery of postischemic stunned myocardium after induced expression of myocardial heat-shock protein (HSP 70). *J. Thorac. Cardiovasc. Surg.* **109,** 753–764.

Shigematsu S, Sato T, Abe T, Saikawa T, Sakata T and Arita M. (1995) Pharmacological evidence for the persistent activation of ATP-sensitive K^+ channels in early phase of reperfusion and its protective role against myocardial stunning. *Circulation* **92,** 2266–2275.

Speechly-Dick ME, Grover GJ and Yellon DM. (1995) Does ischemic preconditioning in the human involve protein kinase C and the ATP-dependent K^+ channel? Studies of contractile function after simulated ischemia in an atrial *in vitro* model. *Circ. Res.* **77,** 1030–1035.

Steenbergen C, Perlman ME, London RE and Murphy E. (1993) Mechanism of preconditioning; ionic alterations. *Circ. Res.* **72,** 112–125.

Subramanian VA, Sani G, Benetti FJ and Calafiore AM. (1995) Minimally invasive coronary bypass surgery: a multi-center report of preliminary clinical experience. *Circulation* **92** (suppl. I), I-645 .

Sugimoto S, Puddu PE, Monti F, Schiariti M, Campa PP and Marino B. (1994) Pretreatment with the adenosine triphosphate-sensitive potassium channel opener nicorandil and improved myocardial protection during high-potassium cardioplegic hypoxia. *J. Thorac. Cardiovasc.* **108,** 455–466.

Sugimoto S, Iwashiro K , Monti F, Dawodu AA, Schiariti M and Puddu PE. (1995a) The risk of myocardial stunning is decreased concentration-dependently by K_{atp} channel activation with nicorandil before high K^+ cardioplegia. *Int. J. Cardiol.* **48,** 11–25.

Sugimoto S, Puddu PE, Monti F, Dawodu AA, del Monte F, Schiariti M, Campa PP and Marino B. (1995b) Activation of ATP-dependent K^+ channels enhances myocardial protection due to cold high potassium cardioplegia: a force–frequency relationship study. *J. Mol. Cell. Cardiol.* **27,** 1867–1881.

Sun JZ, Tang XL, Knowlton AA, Park SW, Qiu Y and Bolli R. (1995) Late preconditioning against myocardial stunning. An endogenous protective mechanism that confers resistance to postischemic dysfunction 24 h after brief ischemia in conscious pigs. *J. Clin. Invest.* **95,** 388–403.

Tang XL, Sun JZ, Park SW, Qiu Y and Bolli R. (1994) Adenosine receptor blockade has no effect on myocardial stunning after repetitive ischemia in conscious pigs. *Circulation* **86** (suppl. I), I-475.

Tomai F, Crea F, Gaspardone A, Versaci F, De Paulis R, Penta de Peppo A, Chiariello L and Gioffré PA. (1994) Ischemic preconditioning during coronary angioplasty is prevented by glibenclamide, a selective ATP-sensitive K^+ channel blocker. *Circulation* **90,** 700–705.

Tsuchida A, Miura T, Miki T and Iimura O. (1992) Role of adenosine receptor activation in myocardial infarct size limitation by ischemic preconditioning. *Cardiovasc. Res.* **26,** 456–461.

Tsuchida A, Thompson R, Olsson RA and Downey JM. (1994) The anti-infarct effect of an adenosine A1-selective agonist is diminished after prolonged infusion as is the cardioprotective effect of ischaemic preconditioning in rabbit heart. *J. Mol. Cell. Cardiol.* **26,** 303–311.

Vuorinen K, Ylitalo K, Peuhkurinen K, Raatikainen P, Ala-Râmi A and Hassinen IE. (1995) Mechanisms of ischemic preconditioning in rat myocardium. Roles of adenosine, cellular energy state, and mitochondrial F1F0-ATPase. *Circulation* **91,** 2810–2818.

Walker DM, Walker JM, Pugsley WB, Pattison CW and Yellon DM. (1995) Preconditioning in isolated superfused human muscle. *J. Mol. Cell. Cardiol.* **27,** 1349–1357.

Wall TM, Sheehy R and Hartman JC. (1994) Role of bradykinin in myocardial preconditioning. *J. Pharm. Exp. Ther.* **270,** 681–689.

Walsh RS, Borges M, Thornton JD, Cohen MV and Downey JM. (1995) Hypoxia preconditions rabbit myocardium by an adenosine receptor-mediated mechanism. *Can. J. Cardiol.* **11,** 141–146.

Winter CB, Mitchell MB, Locke-Winter CR, Harken AH and Banerjee A. (1992) Adenosine-induced cardiac preconditioning is dependent upon α1-adrenoreceptor activation. *Circulation* **86** (suppl. I), I-25.

Wollmering MM, Fullerton DA, Walther JM, Harken AH and Banerjee A. (1994) Preconditioning protects function and viability in human ventricle. *Circulation* **90** (suppl. I), I-477.

Yao Z and Gross GJ. (1993) Role of nitric oxide, muscarinic receptors, and the ATP-sensitive K^+ channel in mediating the effects of acetylcholine to mimic preconditioning in dogs. *Circ. Res.* **73,** 1193–1201 .

Yao Z and Gross GJ. (1994) A comparison of adenosine-induced cardioprotection and ischemic preconditioning in dogs. Efficacy, time course and role of K_{atp} channels. *Circulation* **89,** 1229–1236.

Yellon DM, Alkhulaifi AM and Pugsley WB. (1993) Preconditioning the human myocardium. *The Lancet* **342,** 276–277.

Ytrehus K, Liu Y and Downey JM. (1994) Preconditioning protects ischemic rabbit heart by protein kinase C activation. *Am. J. Physiol.* **266,** H1145–1152.

Zucchi R, Ronca-Testoni S, Yu G, Galbani P, Ronca G and Mariani M. (1995) Postischemic changes in cardiac sarcoplasmic reticulum Ca^{2+} channels. A possible mechanism of ischemic preconditioning. *Circ. Res.* **76,** 1049–1056.

10

Are the benefits of angina pectoris related to ischaemic preconditioning?

Michael S. Marber
Department of Cardiology, The United Medical and Dental Schools of Guy's and St Thomas' Hospitals, The Rayne Institute, London SE1 7EH, UK

1. Introduction

The aims of treatment in angina pectoris are to alleviate symptoms and enhance prognosis. Patients may have angina for many years since the prognosis of ischaemic heart disease is principally determined by left ventricular performance (White *et al.*, 1987) rather than symptom severity (Hultgren and Peduzzi, 1984). The prognosis is therefore enhanced by strategies that minimize the attrition of left ventricular function by preventing (Scandinavian Simvastatin Survival Study Group, 1994; Shepherd *et al.*, 1995), limiting necrosis during (The GUSTO investigators, 1993; ISIS-2 Collaborative Group, 1988), or abating remodelling following (AIRE study investigators, 1993; St John Sutton *et al.*, 1994) acute myocardial infarction. Although it is clear that left ventricular performance may be transiently depressed following an episode of angina (Jeroudi *et al.*, 1994) the long-term influence of angina upon the natural history of ischaemic heart disease is less clear.

Ischaemia: preconditioning and adaptation, edited by M.S. Marber and D.M. Yellon

The purpose of this chapter is to draw attention to recent studies suggesting that in patients with ischaemic heart disease, angina paradoxically improves outcome post-infarction. This chapter will discuss the historical and pathophysiological basis for these benefits and their impact on the management of angina pectoris.

2. The mediation of angina

The discomfort of angina pectoris is mediated by cardiac sympathetic afferent neurones (Meller and Gebhart, 1992). The component or consequence of ischaemia that stimulates these neurones is not known for certain (Meller and Gebhart, 1992). However, interstitial adenosine and bradykinin accumulation during myocardial ischaemia are thought to be important mediators (Meller and Gebhart, 1992). Thus in patients with ischaemic heart disease the discomfort of angina can be precisely mimicked by infusion of adenosine into the diseased coronary artery (Crea *et al.*, 1992) and attenuated by the use of relatively specific adenosine type 1 receptor (A_1) antagonists (Gaspardone *et al.*, 1995).

One can therefore view the pathophysiology of angina pectoris as follows. Usually as a result of a stenosis in an epicardial coronary artery there is a limited flow reserve to an area of myocardium. Thus, in response to exercise, any increase in myocardial work is not accompanied by a corresponding increase in myocardial blood flow. Consequently, tissue oxygen tension falls and mitochodrial respiration is diminished, preventing the recharge of adenosine diphosphate (ADP) to adenosine triphosphate (ATP). Hence, ADP enters a nucleoside degradation pathway leading to the formation and accumulation of adenosine.

3. The mediation of classic ischaemic preconditioning and delayed ischaemic preconditioning

Following a brief period of ischaemia the heart becomes resistant to further ischaemia (Marber *et al.*, 1994a,b). For protection to manifest, the interval between the triggering and subsequent period of ischaemia must be less than 60–120 min (Marber *et al.*, 1994a,b). This form of protection is known as classic ischaemic preconditioning. More recently a number of investigators have noted that in animal studies a delayed more protracted period of protection appears at 24 h, and lasts until 72 h after a short episode of ischaemia (Kuzuya *et al.*, 1993; Marber *et al.*, 1993; Yang *et al.*, 1996). This form of protection is known as delayed ischaemic preconditioning or the second window of protection (SWOP) (Marber *et al.*, 1993), however, some aspects of this phenomenon remain controversial (Tanaka *et al.*, 1994). Since classical and delayed ischaemic preconditioning share the same trigger is there any evidence to suggest that they share the same mediator?

Classical ischaemic preconditioning in both animal and human studies is mediated by the accumulation of interstitial adenosine during the short triggering period of ischaemia (Cohen and Downey, 1993). Adenosine binds to A_1 receptors on cardiac myocytes which results in protein kinase C (PKC) activation via a phospholipase C (or D)-dependent pathway (Cohen and Downey, 1993). In addition there is evidence to suggest that the final downstream anti-ischaemic effect is caused by opening of the ATP-dependent potassium channel (K_{ATP}) during ischaemia (Cohen and Downey, 1993; Gross and Auchampach, 1992; Speechly-Dick *et al.*, 1995).

Delayed ischaemic preconditioning has been less extensively investigated and the only information available is from animal studies. These studies indicate that delayed, like classic ischaemic preconditioning, has a proximal signalling pathway involving adenosine, the A_1 receptor and PKC (Baxter *et al.*, 1994). The final alterations within the myocardium that cause delayed protection are unclear but may involve increases in a range of cytoprotective proteins (Marber *et al.*, 1993, 1995).

Thus angina, classic and delayed ischaemic preconditioning all involve the accumulation of interstitial adenosine acting upon A_1 receptors. Is there any evidence to suggest that, like preconditioning, angina is followed by resistance to further ischaemia?

4. Angina during percutaneous transluminal coronary angioplasty (PTCA)

Angina-like chest pain occurs on inflation of the intracoronary balloon during PTCA. A component of this discomfort is almost certainly secondary to direct mechanical stretch of the vessel containing the target lesion (Tomai *et al.*, 1993). Nonetheless, balloon inflation occludes the vessel lumen, obstructs coronary blood flow and results in ST segment alteration on epicardial and surface electrocardiograms (ECGs) (Deutsch *et al.*, 1990). Therefore, it is also likely that a component of the discomfort is secondary to myocardial ischaemia and represents a form of angina.

PTCA has therefore been used as a model to investigate angina and classical ischaemic preconditioning in the human (Cribier *et al.*, 1992; Gaspardone *et al.*, 1995; Tomai *et al.*, 1994). The results of these investigations (Cribier *et al.*, 1992; Gaspardone *et al.*, 1995; Tomai *et al.*, 1994) confirm that all measures of ischaemia are less on subsequent, than on the first balloon inflation and that this effect is not necessarily due to the appearance of radiographically visible collateral channels (Cribier *et al.*, 1992). In addition these studies suggest that the attenuation of ischaemia on late inflations is adenosine receptor and K_{ATP} channel dependent (Gaspardone *et al.*, 1995; Tomai *et al.*, 1994). Can the findings in this unnatural model of angina be extrapolated to the ambulatory patient with spontaneous angina?

5. Warm-up in angina

'First effort', 'warm-up' or 'first hole' angina describe the ability of some patients to exercise to angina, rest and then continue exertion with few or no symptoms. They may be related to 'walk-through' angina or 'second wind' in angina, which similarly describe a phenomenon where angina sufferers exercise to pain and are then able to continue exertion with few or no symptoms. In contrast with warm-up, however, there is no intervening period of rest. Some of these phenomena have been recognized for centuries (Marber *et al.*, 1994a,b; Heberden, 1785) but are perhaps less publicized now because of effective anti-anginal therapy and investigative techniques. For example treadmill exercise testing does not allow rest and imposes incremental increases in effort which make walk-through less likely. Nevertheless warm-up is a very real effect, with approximately one in five patients reporting compatible symptoms (MacAlpin and Kattus, 1966). In recent studies with objective testing, at least 75% of unselected patients with angina show an enhanced performance on second effort (Jaffe and Quinn, 1980; Okazaki *et al.*, 1993; Williams *et al.*, 1985). In these studies, exercise performance on second effort is on average 125% of that on first effort (Jaffe and Quinn, 1980; Okazaki *et al.*, 1993; Williams *et al.*, 1985), but there is a great deal of individual variability. Not surprisingly those individuals who derive the greatest benefit from first effort are those most likely to provide a history of warm-up (Okazaki *et al.*, 1993).

The pathophysiology of warm-up angina is poorly understood. However, the paradox of patients recovering from an episode of angina being more resistant to further effort-induced angina superficially resembles the paradox that led to the discovery of ischaemic preconditioning.

Traditionally, warm-up angina was ascribed to coronary vasodilatation (Wayne and Graybiel, 1934; Wayne and Laplace, 1933), with perhaps the concomitant opening of collateral vessels to support the ischaemic myocardium. This explanation was accepted despite the fact that arterial vasodilators such as aminophylline had little effect on exercise tolerance (Wayne and Laplace, 1933), and the presence of collaterals on angiography was not a predictor of walk through or warm-up angina (MacAlpin and Kattus, 1966). In addition there were other features to warm-up that were difficult to explain such as the fact that second effort would reproducibly exceed first effort providing the two were separated by a rest period of at least 2–5 min but not more than 30–60 min (Wayne and Graybiel, 1934). These time constraints are consistent with warm-up being a form of ischaemic preconditioning.

The suggestion that warm-up, like ischaemic preconditioning, represents an adaptive phenomenon is further supported by invasive investigation. Williams *et al.* (1985) examined 11 patients with stable angina due to a left anterior descending (LAD) stenosis. No attempt was made to elicit a history of warm-up prior to patient selection. Heart rate was incrementally increased by atrial pacing until angina occurred, patients were then rested for between 5 and 10 min before repeating the pacing protocol. During the second period of pacing, anginal symptoms and ST segment depression were diminished. More interestingly, venous blood samples taken from the distal coronary sinus close to the interventricular vein showed that lactate was still being extracted during the second protocol at a timepoint when on average it was being produced during the first pacing protocol (Williams *et al.*, 1985). This change in myocardial metabolism was mirrored by a reduction in myocardial blood flow and oxygen consumption in the second compared with the first pacing period. This latter observation effectively ruled out enhanced collateral flow as the explanation for warm-up.

The results obtained above are very similar to those reported by Okazaki *et al.* (1993). These investigators studied 13 patients with LAD stenosis by performing two sequential supine exercise tests separated by 15 min of rest. They also found that for an equivalent amount of work, angina, ST segment depression and myocardial oxygen consumption were less on second effort (Okasaki *et al.*, 1993). In addition they demonstrated that although angina occurred at the same myocardial oxygen consumption during each test this was achieved at a higher workload on second effort, suggesting that the heart had become regionally more efficient. They related these findings to an enhanced production of adenosine by the ischaemic LAD territory on second effort (Okasaki *et al.*, 1993). Intuitively, adenosine release, like all other measures of ischaemia, should have been reduced on second effort. However, an identical 'adenosine paradox' occurs in ischaemic preconditioning where the enhanced ischaemic production of adenosine is thought to be secondary to an increase in the activity of the adenosine-forming enzyme 5′-nucleotidase immediately after sublethal ischaemia (Kitakaze *et al.*, 1992).

Clearly the reports by Williams *et al.* (1985) and Okazaki *et al.* (1993) imply that a change occurs in myocardial metabolism that reduces the requirements for substrate on second compared with first effort. One explanation is that this effect is secondary to stunning (of regional contractile function) following the first effort/rapid pacing period which may reduce contractile protein energy requirements. This seems unlikely, however, since in both studies myocardial oxygen requirements returned to baseline after the first rapid pacing/exercise period (Okasaki *et al.*, 1993; Williams *et al.*, 1985). In addition it is not clear whether the metabolic demands of stunned myocardium are less or more than those of healthy tissue (Ohgoshi *et al.*, 1991). Other possibilities such as the redistribution of epi- versus endo-myocardial blood flow from one effort to the next seem inherently unlikely and are not easy to prove or disprove (Okasaki *et al.*, 1993). The most likely explanation is that this represents a form of myocardial adaptation akin to ischaemic preconditioning. To demonstrate conclusively that warm-up and ischaemic preconditioning share the same triggers would require A_1 receptor blockade in the patient.

If warm-up is related to ischaemic preconditioning, what is its relationship to walk-through or second wind in angina? In contrast to the 'no-flow' ischaemia of most experimental ischaemic preconditioning protocols, angina is secondary to ischaemia with flow. In such a situation reperfusion or a period of rest may not be necessary to trigger preconditioning, since ischaemia and reperfusion are coincident. Thus one can speculate that when anginal discomfort and A_1 receptor occupancy have reached a critical level, ischaemic preconditioning is triggered, the myocardium becomes more efficient and angina diminishes or disappears. Such disappearance of pain at a critical point during exercise gives rise to the 'second wind' or 'toter punkt' phenomenon (Wenkebach, 1928). This effect is associated with the resolution of ischaemic ST segment changes (Rizi *et al.*, 1982) and the disappearance of regional dyskinesis (Rizi *et al.*, 1982). Both observations are consistent with an adaptive phenomenon.

6. The first description of ischaemic preconditioning?

If warm-up and walk-through are manifestations of ischaemic preconditioning, then the first description of ischaemic preconditioning would predate the observation of Murry *et al.* by almost exactly 200 years (Heberden, 1785).

The first observations on angina by Heberden are detailed and extremely perceptive. In addition there are certain aspects that merit a more detailed examination as they could, with hindsight and optimism, have been a manifestation of ischaemic preconditioning.

In 1772, Dr Heberden read a letter from a patient to assembled physicians at their Royal College in London. The letter was written by a patient who, realising Dr Heberden's interest in angina, wished to further his knowledge. The letter gives an account of the patient's symptoms. In particular it gives a vivid account of walk-through angina.

> " I have frequently, when in company, borne the pain and continued my pace without indulging it; at which times it has lasted from five to perhaps ten minutes, and then gone off, as well as I can recollect, rather suddenly, as it came."

The patient had left a will in which he instructed his executors that should he die he wished his body to be available to Dr Heberden for dissection. Unfortunately, 3 weeks after receiving the letter Dr Heberden was informed that the patient had died.

> "I was informed, that in the midst of a walk which the patient was taking after dinner, he leaned against a post, and begged a passenger (of a passing carriage) to assist him, by whose help he reached a neighbouring house, where he vomited much; and was bled, but died in less than half an hour."

It is likely from this account that Dr Heberden's correspondent probably suffered an acute myocardial infarct. The reduction in left ventricular performance may well have caused autonomic activation which resulted in the vomiting. Despite attempts to reduce his left ventricular end-diastolic pressure by bleeding, he died!

Heberden on hearing the news of the death procured the help of an "experienced and accurate anatomist, Mr J. Hunter, to open the body".

> "In general the viscera were well formed, and in sound state, with marks of great robustness. The contents of the thorax were examined with peculiar attention, particularly the heart with its vessels and valves, and were all found to be in a natural condition, except some specks of a beginning ossification upon the aorta,"

The calcification of the aorta suggests atherosclerosis and this same disease affecting the coronary vasculature was the probable cause of myocardial infarction and death. Interestingly, there was no macroscopic evidence of myocardial infarction. Could it be that the correspondent's angina, though not protecting him from the contractile deficit of ischaemia, had protected him from infarction?

7. Preinfarction angina

Studies in the prethrombolysis era did not consistently show that preinfarction angina had a favourable influence upon left ventricular function and post-infarction mortality (Kloner and Yellon, 1994). There were numerous difficulties with these studies including the absence of thrombolytic reperfusion and adequate control for the differences that exist between patients with and without preinfarction angina (Kloner and Yellon, 1994). Furthermore, these negative findings are not surprising since it is known from animal studies that reperfusion is a prerequisite for the protection of classic ischaemic preconditioning (Kloner and Yellon, 1994). Recent studies with reperfusion during acute myocardial infarction (Andreotti *et al.*, 1996; Anzia *et al.*, 1994, 1995; Iwasaka *et al.*, 1994; Kloner *et al.*, 1995; Nakagawa *et al.*, 1995; Ottani *et al.*, 1995) suggest a reduced enzyme-derived infarct size (Andreotti *et al.*, 1996; Anzai *et al.*, 1994, 1995; Kloner *et al.*, 1995), better preserved left ventricular function (Iwasaka *et al.*, 1994), a better clinical outcome (Anzia *et al.*, 1994, 1995; Iwasaka *et al.*, 1994; Kloner *et al.*, 1995; Nakagawa *et al.*, 1995; Ottani *et al.*, 1995) and an improved short-term prognosis (Anzai *et al.*, 1994, 1995; Kloner *et al.*, 1995) when angina antecedes infarction (see *Table 1*).

Table 1. Summary of studies of angina preinfarction

	Group size (*n*)		Age (years)		% Male		Time to treatment (h)		No collateral (%)		% Anterior MI		Endpoints[a]
Pre-MI angina	Yes	No	Yes	No	Yes	No	Yes	No	Yes	No	Yes	No	
Iwasaka *et al.* (1994)	28	25	55±12	51±11	71	68	4.7±1.2	4.2±1.3	54	48	100	100	5 weeks post-MI Radionuclide ejection fraction (%) 57±16 vs. 43±15, $p<0.001$ LVESV (ml) 55±34 vs. 90±53, $p<0.01$
Anzai *et al.* (1994)	100	53	62±12	63±16	82	75	51[b]	43[b]	48	64	100	100	In-hospital VT/VF(%) 7 vs. 25, LVF 24 vs. 47 VSD/rupture (%) 1 vs. 17 Mortality (%) 11 vs. 28 (all $p<0.05$)
Ottani *et al.* (1995)[c]	12	13	56±9	62±7	75	77	1.3±0.3	1.4±0.3	100	100	100	100	Peak CKMB (IU/l) 86±66 vs. 192±108, $p<0.01$ Hypokinetic segments (*n*) 5.6±4 vs. 11±7.5, $p<0.05$
Kloner *et al.* (1995)	218	198	58±10	59±11	51	49	3.2±0.1	3.0±0.1	85	78	85	74	In-hospital death (%) 3 vs. 8, $p<0.05$ CHF/shock (%) 1 vs. 7, $p<0.01$ Total CK 119 vs. 154, $p<0.01$ Q wave (%) 57 vs. 69, $p<0.01$

(*continued*)

(Table 1 *continued)*

	Group size (n)		Age (years)		% Male		Time to treatment (h)		No collateral (%)		% Anterior MI		Endpoints[a]
Pre-MI angina	Yes	No	Yes	No	Yes	No	Yes	No	Yes	No	Yes	No	
Nakagawa *et al.* (1995)	22	37	59±11	56±11	77	81	4.7±0.9	3.9±1.3	77	83	100	100	Late minus early ejection fraction (%) 7±10 vs. 0±8, $p<0.001$ Regional wall motion late minus early 0.62±0.55 vs. 0.21±0.47, $p<0.001$
Anzai *et al.* (1995)[d]	65	55	64±10	63±13	83	73	58[b]	68[b]	42	78	0	0	In-hospital VT/VF (%) 3 vs. 13, $p<0.05$ Heart failure (%) 18 vs. 38, $p<0.01$ Mortality (%) 3 vs. 13, $p<0.05$
Andreotti *et al.* (1996)	14	9	59±10	58±8	71	78	3.0±1.2	3.8±1.4	Unknown		36	44	TIMI 3 at 35 min post-thrombolysis (%) 64 vs. 0, $p=0.006$ Integrated CKMB (IU/l 24 h) 1716±1171 vs. 4267±3252, $p=0.009$

[a] For all endpoints data presented as preinfarction angina present vs. preinfarction angina absent.
[b] Expressed as the percentage treated within 6 h of symptoms.
[c] A third group with angina for more than 1 week prior to index infarction is not included.
[d] Includes some of the same patients as in Anzai *et al.* (1994) therefore only data from inferior infarction used in table.
NB. Only studies of preinfarction angina with matched patient characteristics, short-term endpoints and where acute reperfusion is attempted are included in this table.

Patients with angina prior to infarction may differ in a number of respects from patients without. Some of these differences may be sufficient to account for the benefit of prodromal angina without the need to evoke ischaemic preconditioning. For example, patients with angina may present to hospital and have thrombolytic therapy earlier, have a different pattern of coronary artery disease or infarction (Andreotti *et al.*, 1996), be more likely to be taking anti-anginal medication and have better developed collateral support to the infarct-related artery (see below).

The time from symptom onset to thrombolytic therapy or primary angioplasty does not appear to be significantly shorter in patients with prior angina. In fact the reverse is true. In the largest study based upon the TIMI 4 database, thrombolysis in patients with prior angina compared to those without was significantly delayed (Kloner *et al.*, 1995). In addition the other variables that may have influenced outcome did not differ between groups (see *Table 1*). This suggests that preinfarction angina independently predicts a better outcome.

The magnitude of the advantage conferred by preinfarction angina is substantial. For example, the relative risk of death without, compared to with, antecedent angina before anterior infarction can be increased as much as two-fold (Anzai *et al.*, 1995).

The interval between the last episode of angina and the subsequent acute myocardial infarction was carefully recorded in two of the above studies (Nakagawa, 1995; Ottani, 1995). In these studies the median intervals were in excess of 11 h, suggesting that the mechanism for protection corresponds not to classic, but rather to delayed ischaemic preconditioning.

The findings of the most recent of the studies (Andreotti *et al.*, 1996) suggest that an alternative, though not mutually exclusive, mechanism is responsible for the benefit of preinfarction angina. Andreotti *et al.* (1996) have demonstrated convincingly that thrombolytic reperfusion of the infarct related artery is more rapid in those patients with a history of unstable angina. In addition, this observation was accompanied by an approximate 50% reduction in enzymatically derived size of infarction.

Apart from time to reperfusion, collateral support to an infarct related artery is known to be a powerful predictor of an improved outcome (Habib *et al.*, 1991). Surprisingly, a history of preinfarction angina was not associated with an increased incidence of radiographically visible collaterals apart from when angina occurred for more than 1 week prior to infarction (Nakagawa *et al.*, 1995). The difficulty with this aspect of the above studies is that collaterals may only be visible when a pressure gradient exists between two different vascular beds (Schaper, 1967). Hence collaterals present when an infarct related artery is occluded may vanish upon reperfusion. Since, in the studies above (Andreotti *et al.*, 1996; Anzia *et al.*, 1994, 1995; Iwasaka *et al.*, 1994; Kloner *et al.*, 1995; Nakagawa *et al.*, 1995; Ottani *et al.*, 1995), coronary angiography was usually performed at a time after thrombolysis when reperfusion would often already have been achieved, the degree of collateral support during the critical period of infarction was unknown. Despite this theoretical drawback, in the minority of patients with persistent occlusion of the infarct related artery a recent history of previous angina still did not predict greater collateral support (Kloner *et al.*, 1995).

8. Myocardial collaterals

In the above studies (Andreotti *et al.*, 1996; Anzia *et al.*, 1994, 1995; Iwasaka *et al.*, 1994; Kloner *et al.*, 1995; Nakagawa *et al.*, 1995; Ottani *et al.*, 1995), although angina of recent onset was not associated with a greater degree of collateral support to the infarct related artery, in patients with long-standing angina the collateral score was increased (Nakagawa *et al.*, 1995).

Descriptive studies in the human have indicated the benefits of coronary collaterals for over a decade (Gregg and Patterson, 1980). With the advent of trials of thrombolytic therapy involving early angiography, it became apparent that patients with, compared to patients without, radiographically demonstrable collaterals had improved post-infarction left ventricular function (Habib *et al.*, 1991), reduced creatine kinase-derived infarct size (Habib *et al.*, 1991), and less chance of adverse post-infarction remodelling (Hirai *et al.*, 1989) and aneurysm formation (Hirai *et al.*, 1989). In addition, those patients with collaterals had a greater benefit from late thrombolysis (Rentrop *et al.*, 1989). Moreover, other indirect evidence suggested that collaterals may even be able to limit necrosis in patients with failed reperfusion or persistent occlusion of the infarct related artery (Rentrop *et al.*, 1989; Topol and Ellis, 1991).

Clearly, in the setting of an obstruction in an epicardial coronary artery, collateral blood flow is profoundly protective, but what governs the formation of collaterals in the adult heart?

It is now widely accepted that myocardial ischaemia somehow triggers collateral growth (Sasayama and Fujita, 1992). Initial hypotheses evoked the necessity for the creation of a pressure gradient between obstructed and non-obstructed coronary arteries (Schaper, 1967). This gradient provided the potential energy required to open bridging collaterals from the high to the low pressure territory (Schaper, 1967). However, the recognition that collateral formation occurred in anaemic dogs or following microsphere occlusion of the microcirculation (Chilian *et al.,* 1990), where pressure in the epicardial coronary arteries was unaltered, rendered this hypothesis untenable. These and other studies suggested that ischaemia *per se* is a sufficient stimulus for collateral growth, and supported the alternative hypothesis that collaterals grow in response to a chemical signal originating from ischaemic myocardium (Kumar *et al.,* 1983). The observation that heparin-containing mast cells were associated with capillary sprouting and that exogenous heparin enhanced ischaemia-induced collateral formation suggested that this chemical signal may directly or indirectly involve heparin. Ultimately, this information led to the isolation of a number of heparin-binding growth factors (Casscells *et al.,* 1990).

In recent years the genes of putative collateral-forming growth factors have been cloned and recombinant proteins generated. This has recently resulted in the availability of recombinant growth factors which can be administered directly to the heart to test the hypothesis that chemical signals are responsible for collateral growth. In these studies fibroblast growth factors (FGF) and vascular endothelial growth factors (VEGF) have been shown to be angiogenic (Lazarous *et al.,* 1995). In addition the DNA sequences responsible for gene regulation in response to hypoxia have been examined closely. This examination has provided an insight into the local environment within the myocardium which causes growth factor expression.

The indications are that angiogenic growth factor expression is mediated by a fall in local oxygen tension as well as a synergistic interaction between growth factors (Stavri *et al.,* 1995). Thus, upstream of the VEGF gene are short sequences of DNA known to bind a hypoxia inducible factor (Liu *et al.,* 1995; Wang *et al.,* 1995). The binding of this factor is sufficient and necessary for the hypoxic induction of a number of genes, the archetypal example being erythropoietin (Wang *et al.,* 1995). Thus hypoxia within an area of myocardium is likely to be a stimulus for collateral growth.

As already described above, the sensation of angina pectoris is indirectly indicative of myocardial hypoxia since adenosine will only accumulate when the myocardial oxygen tension is sufficiently low to restrict oxidative phosphorylation. Angina pectoris may therefore indicate three separate adaptive processes – classic preconditioning, delayed preconditioning and the substrate for collateral formation.

9. Implications for the management of angina pectoris

The work described above suggests a very close relationship between the discomfort or pain of angina and an enhanced resistance to ischaemia through preconditioning, collateral growth or perhaps more rapid reperfusion (Braunwald, 1996). Many aspects of these relationships are unknown. For example, does silent ischaemia confer a similar advantage? Does suppressing angina abolish the advantage? Is the advantage confined to unstable preinfarction angina?

Clearly, any inferences made on the basis of the above studies are necessarily speculative. Although angina may result in protection there is no doubt that there are also harmful consequences with the occurrence of potentially fatal and non-fatal arrhythmia (Berntsen *et al.,* 1995) and transient disturbances of left ventricular contractile performance (Jeroudi *et al.,* 1994). In addition, as Andreotti *et al.* (1996) and Braunwald (1996) emphasize, the pathophysiology of the plaque that gives rise to unstable angina is different from that causing stable angina. An implication of these differences is that it is more likely for a plaque causing unstable angina to be the site of an infarct-related artery occlusion than for a plaque causing stable angina. Therefore, in stable angina even if preconditioning and collateral growth occur it may be in an area of myocardium remote from subsequent infarction. Hence, in patients with stable angina the risk of an adverse event during symptoms is almost certain to exceed the theoretical benefit following symptoms. Meanwhile, in patients with unstable angina, certain treatments known to reduce symptoms such as β-blockade, heparin and aspirin definitely improve outcome. At present therefore there is insufficient reason to alter anti-anginal prescribing habits or advice to patients because of a fear that symptom reduction in stable (or unstable) angina will prevent preconditioning or collateral growth.

These reservations aside, is there likely to be any advantage of one anti-anginal agent over another? The very remote possibility that anti-anginal therapy may have a detrimental effect makes the use of agents, such as β-blockers, with proven efficacy in secondary prevention even more attractive (Secondary Prevention Panel, 1995). A view that is consistent with current consensus guidelines. In contrast some of the doubts that have been raised over the long-term safety of high doses of short acting calcium antagonists may need to be taken into account (Furberg *et al.*, 1995). Whilst (in Europe and Japan) the newest addition to the anti-angina armamentarium, the K_{ATP} channel openers, may exert their anti-anginal effect through the very mechanisms that underlie classic ischaemic preconditioning. This class of agent may therefore prove the exception to the rule since gain may occur without pain. Conversely the oral sulphonylurea hypoglycaemics glibenclamide and tolbutamide, which block the opening of the K_{ATP} channel, may abolish the advantages of angina, an effect that could contribute to the high mortality of acute myocardial infarction in diabetic patients on oral therapy (Knatterud *et al.*, 1971).

In summary, there is accumulating evidence that in patients with ischaemic heart disease angina pectoris is not all bad. However, at present the benefits are not sufficiently certain to impact upon clinical practice.

References

Acute Infarction Ramipril Efficacy (AIRE) Study Investigators. (1993) Effect of ramipril on mortality and morbidity of survivors of acute myocardial infarction with evidence of heart failure. *Lancet* **342,** 821–828.

Andreotti F, Pasceri V, Hackett DR, Davies GJ, Haider AW and Maseri A. (1996) Preinfarction angina as a predictor of more rapid coronary thrombolysis in patients with acute myocardial infarction. *N. Engl. J. Med.* **334,** 7–12.

Anzai T, Yoshikawa T, Asakura Y, Sumihisa A, Meguro T, Akaisha M, Mitamura H, Handa S and Ogawa S. (1994) Effect on short-term prognosis and left ventricular function of angina pectoris prior to first Q-wave anterior wall acute myocardial infarction. *Am. J. Cardiol.* **74,** 755–759.

Anzai T, Yoshikawa T, Asakura Y, Sumihisa A, Akaisha M, Mitamura H, Handa S and Ogawa S. (1995) Preinfarction angina as a major predictor of left ventricular function and long-term prognosis after a first Q wave myocardial infarction. *J. Am. Coll. Cardiol.* **26,** 319–327.

Baxter GF, Marber MS, Patel VC and Yellon DM. (1994) Adenosine receptor involvement in a delayed phase of myocardial protection 24 hours after ischemic preconditioning. *Circulation* **90,** 2993–3000.

Berntsen RF, Gunnes P and Rasmussen K. (1995) Pattern of coronary artery disease in patients with ventricular tachycardia and fibrillation exposed by exercise-induced ischaemia. *Am. Heart J.* **129,** 733–738.

Braunwald E. (1996) Acute myocardial infarction – the value of being prepared. *N. Engl. J. Med.* **334,** 51–52.

Casscells W, Speir E, Sasse J, *et al.* (1990) Isolation, characterization, and localization of heparin-binding growth factors in the heart. *J. Clin. Invest.* **85,** 433–441.

Chilian WM, Mass HJ, Williams SE, Layne SM, Smith EE and Scheel KW. (1990) Microvascular occlusions promote coronary collateral development in conscious dogs. *Am. J. Physiol.* **258,** H1103–H1111.

Cohen MV and Downey JM. (1993) Ischaemic preconditioning: can the protection be bottled? *Lancet* **342,** 6.

Crea F, Gaspardone A, Kaski JC, Davies G and Maseri A. (1992) Relation between stimulation site of cardiac afferent nerves by adenosine and distribution of cardiac pain: results of a study in patients with stable angina. *J. Am. Coll. Cardiol.* **20,** 1498–1502.
Cribier A, Korsatz L, Koning R, Rath P, Gamra H, Stix G, Merchant S, Chan C and Letac B. (1992) Improved myocardial response and enhanced collateral circulation with long repetitive coronary occlusion during angioplasty. A prospective study. *J. Am. Coll. Cardiol.* **20,** 578–586.
Deutsch E, Berger M, Kussmaul WG, Hirshfeld JW, Herrmann HC and Laskey WK. (1990) Adaptation to ischemia during percutaneous transluminal coronary angioplasty: clinical hemodynamic and metabolic features. *Circulation* **82,** 2044–2051.
Furberg CD, Psaty BM and Meyer JV. (1995) Nifedipine. Dose-related increase in mortality in patients with coronary heart disease. *Circulation* **92,** 1326–1331.
Gaspardone A, Crea F, Tomai F, Versaci F, Iamele M, Gioffrè G, Chiariello L and Gioffrè PA. (1995) Muscular and cardiac adenosine-induced pain is mediated by A_1 receptors. *J. Am. Coll. Cardiol.* **25,** 251–257.
Gregg DE and Patterson RE. (1980) Functional importance of the coronary collaterals. *N. Engl. J. Med.* **303,** 1404–1406.
Gross GJ and Auchampach JA. (1992) Blockade of ATP-sensitive potassium channels prevents myocardial preconditioning in dogs. *Circ. Res.* **70,** 223–233.
GUSTO investigators. (1993) An international randomized trial comparing four thrombolytic strategies for acute myocardial infarction. *N. Engl. J. Med.* **329,** 673–682.
Habib GB, Heibeg J, Forman SA, Brown BG, Roberts R, Terrin ML and Bolli R. (1991) Influence of coronary collateral vessels on myocardial infarct size in humans. Results of phase I thrombolysis in myocardial infarction (TIMI) trial. The TIMI Investigators. *Circulation* **83,** 739–746.
Heberden W. (1785) A letter to Dr Heberden, concerning the angina pectoris; and an account of the dissection of one who had been troubled with that disorder. Read at the college Nov 17, 1772. *Med. Trans. Publ. Coll. Physicians Lond.* **3,** 1–11.
Hirai T, Fujita M, Nakajima H, Asanoi H, Yamanishi K, Ohno A and Sasayama S. (1989) Importance of collateral circulation for prevention of left ventricular aneurysm formation in acute myocardial infarction. *Circulation* **79,** 791–796.
Hultgren HN and Peduzzi P. (1984) Relation of severity of symptoms to prognosis in stable angina pectoris. *Am. J. Cardiol.* **54,** 988–993.
ISIS-2 (Second International Study of Infarct Survival) Collaborative Group. (1988) Randomized trial of IV streptokinase, oral aspirin, both, or neither among 17187 cases of suspected acute myocardial infarction. *Lancet* **ii,** 349–360.
Iwasaka T, Nakamura S, Karakawa M, Sugiura T and Inada M. (1994) Cardioprotective effect of unstable angina to acute anterior myocardial infarction. *Chest* **105,** 57–61.
Jaffe MD and Quinn NK. (1980) Warm-up phenomenon in angina pectoris. *Lancet* **2,** 934–936.
Jeroudi MO, Cheirif J, Habib G and Bolli R. (1994) Prolonged wall motion abnormalities after chest pain at rest in patients with unstable angina: a possible manifestation of myocardial stunning. *Am. Heart J.* **127,** 1241–1250.
Kitakaze M, Hori M, Takashima S, Sato H, Inoue M and Kamada T. (1992) Ischemic preconditioning increases adenosine release and 5′-nucleotidase activity during myocardial ischaemia and reperfusion in dogs. *Circulation* **87,** 208–215.
Kloner RA and Yellon DM. (1994) Does ischemic preconditioning occur in patients? *J. Am. Coll. Cardiol.* **24,** 1133–1142.
Kloner RA, Shook T, Pryzklenk K, *et al.* (1995) Previous angina alters in-hospital outcome in TIMI-4. A clinical correlate to preconditioning? *Circulation* **91,** 37–47.
Knatterud GL, Meinert CL, Klimt CR, Osbourne RK and Martin DB. (1971) (The University Group Diabetes Program) Effects of hypoglycemic agents on vascular complication in patients with adult-onset diabetes. *JAMA* **217,** 777–784.
Kumar S, West D, Shahabuddin S, Arnold F, Haboubi N, Reid H and Carr T. (1983) Angiogenesis factor from human myocardial infarcts. *Lancet* **2,** 364–367.
Kuzuya T, Hoshida S, Yamashita N, Fuji H, Oe H, Hori M, Kamada T and Tada M. (1993) Delayed effects of sublethal ischemia on the acquisition of tolerance to ischemia. *Circ. Res.* **72,** 1293–1299.

Lazarous DF, Scheinowitz M, Shou M, *et al*. (1995) Effects of chronic systemic administration of basic fibroblast growth factor on collateral development in the canine heart. *Circulation* **91,** 145–153.

Liu Y, Cox SR, Morita T and Kourembanas S. (1995) Hypoxia regulates vascular endothelial growth factor gene expression in endothelial cells. Identification of a 5′ enhancer. *Circ. Res.* **77,** 638–643.

MacAlpin RN and Kattus AA. (1966) Adaptation to exercise in angina pectoris. *Circulation* **33,** 183–201.

Marber MS, Latchman DS, Walker JM and Yellon DM. (1993) Cardiac stress protein elevation 24 hours following brief ischemia or heat stress is associated with resistance to infarction. *Circulation* **88,** 1264–1272.

Marber MS, Joy MD and Yellon DM. (1994) Warm-up in angina: is it ischaemic preconditioning? *Br. Heart J.* **72,** 213–215.

Marber MS, Walker DM and Yellon DM. (1994) Ischaemic preconditioning. *Br. Med. J.* **308,** 1–2.

Marber MS, Mestril R, Chi S-H, Sayen MR, Yellon DM and Dillmann WG. (1995) Overexpression of the rat inducible 70 kiloDalton heat stress protein in a transgenic mouse increases the resistance of the heart to ischemic injury. *J. Clin. Invest.* **95,** 1446–1456.

Meller ST and Gebhart GF. (1992) A critical review of the afferent pathways and the potential chemical mediators involved in cardiac pain. *Neuroscience* **48,** 501–524.

Nakagawa Y, Ito H, Kitakaze M, Kusuoka H, Hori M, Kuzuya T, Higashino Y, Fujii K and Minamino T. (1995) Effect of angina pectoris on myocardial protection in patients with reperfused anterior wall myocardial infarction: retrospective clinical evidence of preconditioning. *J. Am. Coll. Cardiol.* **25,** 1076–1083.

Ohgoshi Y, Goto Y, Futaki S, Yaku H, Kawaguchi O and Suga H. (1991) Increased oxygen cost of contractility in stunned myocardium of dog. *Circ. Res.* **69,** 975–988.

Okazaki Y, Kodama K, Sato H, Kitakaze M, Hirayama A, Mishima M, Hori M and Inoue M. (1993) Attenuation of increased regional myocardial oxygen consumption during exercise as a major cause of warm-up phenomenon. *J. Am. Coll. Cardiol.* **21,** 1597–1604.

Ottani F, Galvani M, Ferrini D, Sorbello F, Limonetti P, Pantioli D and Rusticali F. (1995) Prodromal angina limits infarct size. A role for ischemic preconditioning. *Circulation* **91,** 291–297.

Rentrop KP, Feit R, Sherman W, *et al*. (1989) Late thrombolytic therapy preserves left ventricular function in patients with collateralized total coronary occlusion: primary endpoint findings of the Second Mount Sinai-New York University Reperfusion Trial. *J. Am. Coll. Cardiol.* **14,** 58–64.

Rizi HR, Kline RC, Besozzi MC, Kalf V, Rabinovitch M, Chan W, Thrall JH and Pitt B. (1982) Walk-through angina phenomenon demonstrated by graded exercise radionuclide ventriculography: possible coronary spasm mechanisms. *Am. Heart J.* **103,** 292–294.

Sasayama S and Fujita M. (1992) Recent insights into coronary collateral circulation. *Circulation* **85,** 1197–1204.

Scandinavian Simvastatin Survival Study Group. (1994) Randomised trial of cholesterol lowering in 4444 patients with coronary heart disease: the Scandinavian Simvastatin Survival Study (4S) *Lancet* **344,** 1383–1389.

Schaper W. (1967) Tangential wall stress as a moulding force in the development of collateral vessels in the canine heart. *Experientia* **23,** 595–596.

Secondary Prevention Panel. (1995) Preventing heart attack and death in patients with coronary disease. *Circulation* **92,** 2–4.

Shepherd J, Cobbe SM, Ford I, Isles CG, Lorimer AR, MacFarlane PW, McKillop JH and Packard CJ. (1995) Prevention of coronary heart disease with pravastatin in men with hypercholesterolemia. *N. Engl. J. Med.* **333,** 1301–1307.

Speechly-Dick ME, Grover GJ and Yellon DM. (1995) Does ischemic preconditioning in the human involve protein kinase C and the ATP-dependent potassium channel. *Circ. Res.* **77,** 1030–1035.

Stavri GT, Zachary IC, Baskerville PA, Martin JF and Erusalimsky JD. (1995) Basic fibroblast growth factor upregulates the expression of vascular endothelial growth factor in vascular smooth muscle cells. Synergistic interaction with hypoxia. *Circulation* **92,** 11–14.

St John Sutton M, Pfeffer MA, Plappert T, *et al.* (1994) Quantitative two-dimensional echocardiographic measurements are major predictors of adverse cardiovascular events after acute myocardial infarction. The protective effect of captopril. *Circulation* **89,** 68 –75.

Tanaka M, Fujiwara H, Yamasaki K, Miyamae M, Yokota R, Hasegawa K, Fujiwara T and Sasyama S. (1994) Ischemic preconditioning elevates cardiac stress protein but does not limit infarct size 24 or 48 h later in rabbits. *Am. J. Physiol.* **267,** H1476–H1482.

Tomai F, Crea F, Gaspardone A, Versaci F, Esposito C, Chiarello L and Gioffre PA. (1993) Mechanisms of cardiac induced pain during coronary angioplasty. *J. Am. Coll. Cardiol.* **22,** 1892–1896.

Tomai F, Crea F, Gaspardone A, *et al.* (1994) Ischemic preconditioning during coronary angioplasty is prevented by glibenclamide, a selective ATP-sensitive K^+ channel blocker. *Circulation* **90,** 700–705.

Topol EJ and Ellis SG. (1991) Coronary collaterals revisited. Accessory pathway to myocardial preservation during infarction. *Circulation* **83,** 1084–1086.

Wang GL, Jiang B-H, Rue EA and Semenza GL. (1995) Hypoxia-inducible factor 1 is a basic–helix–loop–helix–PAS heterodimer regulated by cellular O_2 tension. *Proc. Natl Acad. Sci. USA* **92,** 5510–5514.

Wayne EJ and Graybiel A. (1934) Observations on the effect of food, gastric distension, external temperature, and repeated exercise on angina of effort, with a note on angina sine dolore. *Clin. Sci.* **1,** 287–304.

Wayne EJ and Laplace LB. (1933) Observations on angina of effort. *Clin. Sci.* **1,** 103–129.

Wenkebach KF. (1928) Toter Punkt, second wind und angina pectoris. *Wien. Klin. Wochenschr.* **41,** 1–10.

White HD, Norris RM, Brown MA, Brandt PWT, Whitlock RML and Wild CJ. (1987) Left ventricular end-systolic volume as the major determinant of survival after recovery from myocardial infarction. *Circulation* **76,** 44–51.

Williams DO, Bass TA, Gewirtz H and Most AS. (1985) Adaptation to the stress of tachycardia in patients with coronary artery disease: insight into the mechanism of the warm-up phenomenon. *Circulation* **71,** 687–692.

Yang XM, Baxter GF, Heads RJ, Yellon DM, Downey JM and Cohen MV. (1996) Infarct limitation in a second window of protection in a conscious rabbit model. *Cardiovasc. Res.,* in press.

11

A clinical overview of preconditioning

Robert A. Kloner
Heart Institute Research Laboratory, Good Samaritan Hospital, University of Southern California, 1225 Wilshire Boulevard, Los Angeles, CA 90017, USA

1. Introduction

Over the last several years data have been emerging which support the concept that preconditioning occurs in the human heart (Kloner and Yellon, 1994). Data from human myocardial biopsies obtained at the time of coronary artery bypass grafting (Yellon *et al.*, 1993), data from *in vitro* studies of right atrial trabeculae (Walker *et al.*, 1995) and isolated human cardiac myocytes (Ikonomodis *et al.*, 1994a) have shown that human tissue can demonstrate many of the features of preconditioned myocardium observed in experimental studies. There are also data that now support the concept that brief episodes of myocardial ischaemia in man, induced by angioplasty, by exercise or pacing-induced ischaemia, or preinfarct angina, may precondition the heart. The purpose of the following review is to describe the evidence that angina (induced by angioplasty, exercise or pacing, or preinfarct angina) may result in preconditioning in man.

2. Angina induced by angioplasty

Perhaps one of the great advantages of studying patients undergoing angioplasty is that the technique allows us to determine the

Ischaemia: preconditioning and adaptation, edited by M.S. Marber and D.M. Yellon

consequences of a brief controlled coronary artery occlusion on the human heart. While the angioplasty balloon is inflated in a patient's coronary artery, there are several consequences: the patient often experiences angina pectoris, the electrocardiogram demonstrates ST-segment elevation, lactate is produced and ventricular dysfunction may occur. Deutsch *et al.* (1990) were among the first to make the observation that there was less chest pain, less ST-segment elevation and less lactate production on a second 90 sec left anterior descending (LAD) balloon inflation, compared to a first inflation. Importantly, in this study there was no increase in great cardiac vein flow during the second inflation compared to the first, suggesting that the beneficial effects of the second inflation were not due to acute recruitment of extensive collateral flow. Cribier *et al.* (1992) studied patients undergoing five sequential angioplasty balloon inflations and also described less chest pain and ST-segment shift with subsequent balloon inflations compared to a first inflation. In 10 of 17 patients there was angiographic evidence of recruitment of collaterals; in the remaining seven patients there was not. Yet in these seven patients there was still less chest pain and ST-segment shift. Hence these later patients may have experienced true ischaemic preconditioning unrelated to recruitment of collateral vessels. Several experimental studies have linked preconditioning to activation of the adenosine triphosphate (ATP)-sensitive potassium (K_{ATP}) channel and adenosine receptors (Ikonomidis *et al.*, 1994b; Gross *et al.*, 1994). There is now evidence that these same mechanisms may play a role in patients undergoing sequential balloon inflations during angioplasty. Tomai *et al.* (1994a) observed that glibenclamide (a blocker of K_{ATP}) prevented the beneficial effects of a second balloon inflation on ST-segment shift. Recently, two adenosine receptor antagonists aminophylline (Claeys *et al.*, 1994) and bamiphylline (Tomai *et al.*, 1994b) were also shown to block the beneficial effects of repetitive angioplasty.

A practical consideration of these observations is that difficult or high-risk angioplasties in which the occluded vessel supplies a large 'area at risk' might benefit from a staged procedure whereby the balloon is initially inflated for 60–90 sec, reperfusion allowed for several minutes, and then second, third and fourth inflations are produced. The 'ischaemic burden' during the subsequent inflation compared to the first should be lessened by preconditioning, and/or recruitment of collateral vessels or both. Angioplasty might also serve as a useful tool in testing 'preconditioning mimetic agents' – those agents which can stimulate the pathways that lead to preconditioning without inducing ischaemia.

Examples of such agents include adenosine receptor agonists such as *R*(–)-*N*-(2-phenylisopropyl)-adenosine (PIA) (Hale *et al.*, 1993), and the K_{ATP} channel opener Nicoradil (Gross *et al.*, 1994). If such agents are going to be applied to clinical cardiovascular disease, they could be first tested in an angioplasty protocol to determine whether they are effective in man. For example, patients undergoing proximal LAD angioplasty could be randomized to a placebo control or preconditioning-mimetic agent prior to undergoing a first balloon inflation. A reduction in chest pain, ST-segment elevation and lactate production in the treated group would suggest that the preconditioning-mimetic was truly anti-ischaemic.

3. Angina induced by an increase in oxygen demand

Clinicians have long known of patients with angina who describe the so-called 'warm-up phenomena'. The typical history is that the patient is walking and then develops angina pectoris. The patient stops, rests and then a few minutes later is able to continue his/her walk without any further symptoms. This phenomenon may represent ischaemic preconditioning. It has also been described in patients undergoing serial exercise stress tests with a several minute rest period between tests. Typically patients are able to exercise longer and develop less electrocardiogram (ECG) manifestations of ischaemia on a second compared to a first test (Jaffe and Quinn, 1980; MacAlpin and Kattus, 1966; Marber *et al.*, 1994; Maybaum *et al.*, 1995; Okazaki *et al.*, 1993). One study (Okazaki *et al.*, 1993) showed that this phenomenon was not associated with an increase in great cardiac vein flow, suggesting that there was no acute recruitment of collateral vessels. They did, however, observe a decrease in myocardial oxygen demand on the second exercise test compared to the first. A similar observation was reported by Williams *et al.* (1985). They studied patients in the catheterization lab who were subjected to rapid pacing, in order to induce ischaemia. Patients experienced two sequential pacing periods, with a rest period in between. During the second pacing period, there was less angina, less ST-segment shift and less lactate production compared to the first without an increase in peak coronary artery blood flow and with a decrease in myocardial oxygen demand.

A practical implication of these observations is that patients with coronary artery disease who are participating in exercise programmes may benefit from a brief warm-up period prior to heavier physical exertion.

4. Angina prior to myocardial infarction as a source of preconditioning ischaemia

We have observed in the TIMI 4 study that about half of patients have histories of angina prior to myocardial infarction (Kloner *et al.*, 1995). Those that had histories of angina had smaller infarct sizes as assessed by creatine kinase (CK) curves and development of Q waves on the ECG, compared to patients without histories of previous angina (Kloner *et al.*, 1995). In addition, prior angina also appeared to confer additional benefits on patients with myocardial infarction: lower hospital mortality and less congestive heart failure and shock. These benefits were observed despite the fact that patients with previous angina were more likely to have multivessel coronary artery disease and had a longer period of time from the onset of chest pain to the administration of thrombolytic therapy. (We surmised that patients with a history of angina who then develop the chest pain of myocardial infarction may delay seeking help because they are unsure of whether their chest pain is just their usual angina or a more serious event.) The benefits of prior angina were observed in patients with a history of angina as well as patients with angina within 48 h of the index infarction; the benefits were independent of the presence or absence of large epicardial coronary artery collaterals visualized on angiography as well as the use of anti-anginal drugs and aspirin. A potential mechanism of the beneficial affects of prior angina includes the concept that angina, a form of brief ischaemia, preconditioned the heart prior to a myocardial infarction. It is not possible to totally rule out the development of small intramural collaterals (which cannot be visualized on coronary angiography) among patients who have histories of prior angina, as an alternative explanation. Nevertheless, there have been several reports suggesting that a history of angina confers benefits upon patients who subsequently develop myocardial infarction. These benefits have included not only reduction in myocardial infarct size assessed enzymatically (Anzai *et al.*, 1994; Kloner *et al.*, 1995; Ottani *et al.*, 1995), but also reductions in mortality (Anzai *et al.*, 1994, 1995; Kloner *et al.*, 1995), improvements in regional wall motion and left ventricular ejection fraction with less pump failure (Anzai *et al.*, 1994, 1995), less incidence of myocardial rupture (Anzai *et al.*, 1994), and in one study less reocclusion (Muller *et al.*, 1990).

Experimental studies have suggested that for classic preconditioning to occur, the episodes of brief ischaemia must occur well within 1–2 h of the prolonged episodes of ischaemia (Li *et al.*, 1992). If the prolonged episode of ischaemia is delayed beyond this time period, then the brief ischaemic preconditioning episodes will lose their ability to reduce myocardial infarct size. However, Yellon and coworkers have observed that a second window of protection occurs between 24 and 72 h after the episodes of brief ischaemic preconditioning (Marber *et al.*, 1993). That is, inducing a prolonged episode of ischaemia 1–3 days after a series of brief episodes of ischaemia once again will be associated with a small infarct. This second window of protection may be more relevant than so-called classic preconditioning when referring to preinfarct angina studies, since in one study, the most common time for angina to occur was about 11 h prior to infarction (Ottani *et al.*, 1995), and in another study angina was common at 25 h before infarction (Nakagawa *et al.*, 1995); yet in both studies, angina conferred a benefit. This time period would fall outside the window of classic preconditioning (within 1–2 h of infarction). The mechanism of this second window of protection may differ from the mechanism of classic preconditioning and possibly involve the activation of heat shock proteins.

Thus, a number of studies have now shown that a history of angina prior to myocardial infarction confers several benefits, including reducing infarct size. This phenomenon is consistent with preconditioning. If preconditioning-mimetic drugs can be developed, then it may be possible to administer such agents to high risk patients or patients with unstable angina and perhaps blunt threatened myocardial infarctions. A practical clinical application of this data is the concept that patients with myocardial infarction who lack a history of prior angina may be at higher risk for larger infarcts, in-hospital mortality and heart failure.

In summary, there is now strong evidence in the clinical literature supporting the concept that preconditioning can occur in patients. This includes less chest pain, ST-segment shift and lactate production on subsequent occlusions, compared to a first coronary occlusion during angioplasty balloon inflations, the angina pectoris 'warm-up' phenomenon and the benefits of preinfarct angina on myocardial infarct size and other in-hospital outcomes. The challenge for the future will be to develop 'preconditioning-mimetic' drugs which stimulate the biochemical pathways of preconditioning, but do not cause angina.

References

Alkhulaifi AM, Pugsley WB and Yellon DM. (1993) The influence of the time period between ischemic preconditioning and prolonged ischemia on myocardial protection. *Cardioscience* **4,** 163–169.

Anzai T, Yoshikawa T, Asakura Y, Abe S, Meguro T, Akaishi M, Miatamura H, Handa S and Ogawa S. (1994) Effect on short-term prognosis and left ventricular function of angina pectoris prior to first Q-wave anterior wall acute myocardial infarction. *Am. J. Cardiol.* **74,** 755–759.

Anzai T, Yoshikawa T, Asakura Y, Abe S, Akaishi M, Mitamura H, Handa S and Ogawa S. (1995) Preinfarction angina as a major predictor of left ventricular function and long-term prognosis after a first Q wave myocardial infarction. *J. Am. Coll. Cardiol.* **26,** 319–327.

Claeys MJ, Vrints CJ, Bosmans JM, Conraads VM and Snoeck JP. (1994) Aminophylline abolishes ischemic preconditioning during angioplasty. *Circulation* (suppl. I, part 2), I-477 (abstract).

Cribier A, Korsatz L, Koning R, *et al.* (1992) Improved myocardial ischemic response and enhanced collateral circulation with long repetitive coronary occlusion during angioplasty: a prospective study. *J. Am. Coll. Cardiol.* **20,** 578–586.

Deutsch E, Berger M, Kussmaul WG, Hirshfeld JW, Hermann HC and Laskey WL. (1990) Adaption to ischemia during percutaneous transluminal coronary angioplasty. Clinical, hemodynamic, and metabolic features. *Circulation* **82,** 2044–2051.

Gross GJ, Yao Z and Auchampach JA. (1994) Role of ATP-sensitive potassium channels in ischemic preconditioning. In: *Ischemic Preconditioning: the Concept of Endogenous Cardioprotection* (eds K Przyklenk, RA Kloner and DM Yellon). Kluwer Academic Publishers, Norwell, MA.

Hale SL, Bellows SD, Hammerman H and Kloner RA. (1993) An adenosine A_1 receptor agonist, *R*(–)-*N*-(2-phenylisopropyl)-adenosine (PIA), but not adenosine itself, acts as a therapeutic preconditioning-mimetic agent in rabbits. *Cardiovasc. Res.* **27,** 2140–2145.

Ikonomidis JS, Tumiati LC, Weisel RD, Mickle DAG and Li R-K. (1994a) Preconditioning human ventricular cardiomyocytes with brief periods of simulated ischaemia. *Cardiovasc. Res.* **28,** 1285–1291.

Ikonomidis JS, Shirai T, Weisel RD and Mickle DAG. (1994b) Human cardiomyocyte preconditioning is adenosine A_1 receptor dependent. *Circulation* **90** (suppl. I, Part 2), I-477.

Jaffe MD and Quinn NK. (1980) Warm-up phenomenon in angina pectoris. *Lancet* **2,** 934–936.

Kloner RA and Yellon D. (1994) Does ischemic preconditioning occur in patients? *Am. Coll. Cardiol.* **24,** 1133–1142.

Kloner RA, Shook T, Przyklenk K, *et al.* (1995) Previous angina alters in-hospital outcome in TIMI 4: a clinical correlate to preconditioning? *Circulation* **91,** 37–45.

Li Y, Whittaker P and Kloner RA. (1992) The transient nature of the effect of ischemic preconditioning on myocardial infarct size and ventricular arrhythmias. *Am. Heart J.* **123,** 346–353.

MacAlpin RN and Kattus AA. (1986) Adaptation to exercise in angina pectoris: the electrocardiogram during treadmill walking and coronary angiographic findings. *Circulation* **33,** 183–201.

Marber MS, Latchman DS, Walker JM and Yellon DM. (1993) Cardiac stress protein elevation 24 hours following brief ischaemia or heat stress is associated with resistance to infarction. *Circulation,* **88,** 1264–1272.

Marber MS, Joy MD and Yellon DM. (1994) Warm-up in angina: is it ischemic preconditioning? *Br. Heart J.* **72,** 213–215.

Maybaum S, Ilan M, Mogilevski J and Tzivoni D. (1995) Repeated exercise induced ischemia causes myocardial preconditioning. *J. Am. Coll. Card.* (suppl.), 22A.

Muller DWM, Topol EJ, Califf RM, *et al.* and TAMI Study Group. (1990) Relationship between antecedent angina pectoris and short term prognosis after thrombolytic therapy for acute myocardial infarction. *Am. Heart J.* **119,** 224–231.

Nakagawa Y, Ito H, Kitakaze M, Kusuoka H, Hori M, Kuzuya T, Higashino Y, Fujii K and Minamino T. (1995) Effects of angina pectoris on myocardial protection in patients with reperfused anterior wall myocardial infarction: retrospective clinical evidence of 'preconditioning'. *J. Am. Coll. Cardiol.* **25,** 1076–1083.

Okazaki Y, Kodama K, Sato H, *et al.* (1993) Attenuation of increased regional myocardial oxygen consumption during exercise as a major cause of warm-up phenomenon. *J. Am. Coll. Cardiol.* **21,** 1597–1604.

Ottani F, Galvani M, Ferrini D, Sorbello F, Limonetti P, Pantoli D and Rusticali F. (1995) Prodromal angina limits infarct size. A role for ischemic preconditioning. *Circulation* **91,** 291–297.

Tomai F, Crea F, Gaspardone A, *et al.* (1994a) Ischemic preconditioning during coronary angioplasty is prevented by glibenclamide, a selective ATP-sensitive K^+ channel blocker. *Circulation* **90,** 700–705.

Tomai F, Crea F, Gaspardone A, Versaci F, DePaulis R, Pento de Peppo A, Chiariello L and Gioffre PA. (1994b) Blockade of A_1-adenosine receptors prevents myocardial preconditioning in man. *Eur. Heart J.* **15** (suppl.), 553.

Walker DM, Walker JM, Pugsley WB, Pattison CW and Yellon DM. (1995) Preconditioning in isolated superfused human muscle. *J. Mol. Coll. Cardiol.* **27,** 1349–1357.

Williams DO, Bass TA, Gerwitz H, *et al.* (1985) Adaptation to the stress of tachycardia in patients with coronary artery disease: insight into the mechanism of the warm-up phenomenon. *Circulation* **71,** 687–692.

Yellon DM, Alkhulaif AM and Pugsley WB. (1993) Preconditioning the human myocardium. *Lancet* **342,** 276–277.

Index

ORDERING DETAILS

Main address for orders

BIOS Scientific Publishers Ltd
9 Newtec Place, Magdalen Road,
Oxford OX4 1RE, UK
Tel: +44 1865 726286
Fax: +44 1865 246823

Australia and New Zealand
DA Information Services
648 Whitehorse Road, Mitcham, Victoria 3132, Australia
Tel: (03) 9210 7777
Fax: (03) 9210 7788

India
Viva Books Private Ltd
4325/3 Ansari Road, Daryaganj, New Delhi 110 002, India
Tel: 11 3283121
Fax: 11 3267224

Singapore and South East Asia
(Brunei, Hong Kong, Indonesia, Korea, Malaysia, the Philippines, Singapore, Taiwan, and Thailand)
Toppan Company (S) PTE Ltd
38 Liu Fang Road, Jurong, Singapore 2262
Tel: (265) 6666
Fax: (261) 7875

USA and Canada
BIOS Scientific Publishers
PO Box 605, Herndon, VA 20172-0605, USA
Tel: (703) 661 1500
Fax: (703) 661 1501

Payment can be made by cheque or credit card (Visa/Mastercard, quoting number and expiry date). Alternatively, a *pro forma* invoice can be sent.

Prepaid orders must include £2.50/US$5.00 to cover postage and packing (two or more books sent post free)